Ministry of Agriculture, Fish
07678901

641.4
AGRICULTURE, FISHERIES A
800437

KW-020-520

Refrigerated storage of
fruit and vegetables

BATH COLLEGE OF HIGHER EDUCATION
NEWTON PARK LIBRARY

DISCARD

Stock No.	Class No.
800437	641.4

London: Her Majesty's Stationery Office

B.C.H.E. – LIBRARY

00021781

© *Crown copyright 1979*
First published 1979

ISBN 0 11 240324 7

Contents

Contents—*continued* *Page*

Foreword

Bulletin 159 published in 1965, was concerned solely with the refrigerated storage of fruit. This new book has been completely revised and extended to include the refrigerated storage of both fruit and vegetables.

The purpose of the book is to help those who either wish to invest in refrigeration equipment and a programme of fruit or vegetable storage, or those who, having existing buildings and equipment, wish to improve their efficiency. General advice on construction and equipment is given in this book and further advice is available from the ADAS Land Service and Farm Mechanisation Advisers.

Specialist advisers have written chapters on crop production, diseases and disorders in relation to storage, further general advice on the growing of crops and the special requirements of those intended for storage can be obtained from local ADAS Horticulture Advisers.

The working party responsible for the revision was chaired by John D Whitwell, and included J P Harrison (Layout, Design and Construction of Stores), D I Bartlett (Plant and Equipment, Technical and Design Section), R F Clements (Storage of Fruit), Dr M R Shipway and A F Crisp (Storage of Vegetables), D Wiggell (Storage Diseases and Disorders of Fruit) and Dr D M Derbyshire (Storage Diseases and Disorders of Vegetables). The helpful co-operation of other ADAS Science Service Specialists, East Malling Research Station and the Food Research Institute is also acknowledged.

G C Williams

Ministry of Agriculture, Fisheries and Food
October 1978

Senior Horticultural Adviser
Agricultural Development and
Advisory Service

The storage of fruit

The object of all fruit storage techniques is to delay ripening so that fruit can be marketed at the optimum time without loss of quality. In essence successful storage depends on efficient control of temperature and humidity and, for appropriate crops, modification of the gaseous composition of the store atmosphere.

In this section storage is taken to include all techniques for the removal of field heat from produce before marketing. Storage may thus vary from the very short term, in which soft fruit is held overnight in simple cold stores, to the long term holding of apples and pears in refrigerated chambers with modified atmospheres.

The nature of the produce determines the length of time it can be stored. Soft fruits cannot be cold stored in air for more than a few days so that storage requirements in terms of capital and equipment are relatively small. Nutrition appears to be of minor importance in the storage performance of soft fruit, but control of rotting by applications of fungicides is a prerequisite for successful storage.

The effect of orchard conditions on apples and pears and their interactions with conditions within stores profoundly affects the storage life and quality of these fruits. The elucidation of the factors affecting storage is the subject of intense research. It is clear that their relative importance increases as the period of storage is prolonged. Whereas apples and pears may be held for a few weeks in cold store with little regard for their field history, full exploitation of their storage potential depends on understanding the effect of the orchard environment on the fruit and providing storage facilities capable of meeting the varied demands upon it. In practice this means highly capitalised controlled atmosphere (CA) storage.

Apples and pears

ASPECTS OF STORAGE

Cold storage in air

Storage in a freely ventilated chamber cooled by means of refrigerating plant is commonly known as cold storage. The principles are simple. Temperature affects all living processes such as growth, ripening, and the progress of rotting. These

1

processes are rapid at high temperatures of 15–21°C and slow at temperatures of 2–7°C. The changes which apples and pears undergo (for example the change in colour from green to yellow, and the softening and sweetening of tissue) are living processes and, as such, are slowed down at low temperatures. Rotting, which is caused by fungi growing in the tissues, is similarly delayed. The temperature of storage, however, must not be too low. Fruit is killed by freezing and when taken from store becomes brown and soggy and is attacked by moulds. Exposure to temperatures approaching freezing point may also injure some fruits. Since varieties differ in their susceptibility to low temperature injury they need to be kept at different temperatures. It must be realised that even when kept under the best conditions there is a limit to the storage life of fruit.

Controlled atmosphere (CA) storage

In CA storage, not only is the temperature controlled but the ventilation of the store is so restricted as to permit the carbon dioxide produced by the respiring fruit to accumulate in the storage atmosphere. This, together with the fall in concentration of oxygen which accompanies it, has the effect of delaying ripening changes. To obtain this accumulation of carbon dioxide the store has to be gas proof; a gas store is, therefore, a cold store which is rendered gas tight and fitted with a ventilating device, by which the rise in concentration of carbon dioxide and the fall in the concentration of oxygen can be regulated.

Details of the design and construction of refrigerated stores for apples and pears in air and CA are given on pages 66 and 92.

Temperature and storage conditions

Without the use of refrigeration apples can rarely be kept in good condition beyond November. With refrigeration the storage period can be extended until December (or January for some varieties) and with CA storage until March or even later depending on the variety (see Table 1). The recommended conditions for storage of the main varieties of apples and pears in air and CA have been determined by experiments over many years and confirmed in commercial practice. Minor modifications which take into account seasonal weather conditions and the nutritional status of the fruit may be advised for particular seasons and for particular orchards. Information on these matters may be obtained from ADAS advisers. Most of the storage work on less widely grown varieties is on a relatively small scale. New varieties of commercial significance are being subjected to further testing at the Ditton Laboratory of East Malling Research Station and ADAS Experimental Horticulture Stations.

CULTURAL FACTORS AFFECTING STORAGE

These can be considered under three headings:

   seasonal weather conditions;

   permanent orchard factors such as soil type, rootstock and age of tree;

   variable factors such as soil management, irrigation, spraying programme
      and fertiliser practice.

2

Table 1 Recommended Storage Conditions for the main varieties of apple (October 1977)

(The suggested terminal date for each set of conditions allows for a further two weeks at ambient temperature to cover grading and distribution)

Variety	Air Storage		Controlled atmosphere (CA)						Notes
	Temp. °C	Terminate	Temp. °C	No scrubber		Using a scrubber			
				% CO_2	Terminate	% CO_2	% O_2	Terminate	
Bramley's Seedling	3·0-4·0	January	3·5-4·5	8-10	May	—	—	—	1 (a and b)
Cox's Orange Pippin and its sports	3·0-3·5	Mid Dec.	3·5-4·0	5	Early Jan.	5	3	February	2 (a and b)
			3·5-4·0			<1	2	Late March	2 (a and b)
Crispin	1·5-2·0	January	3·5-4·0	8	April	—	—	—	1 (a)
Egremont Russet	3·0-3·5	December	3·0-3·5	6-8	January	5	3	Early March	3
Laxton's Superb	1·5-3·5	February	3·5-4·5	6-8	March	7-8	3	Late March	3
Worcester Pearmain	0-1·0	December	0·5-1·0	7-8	February	5	3	Mid March	3

Notes

1 (a) Use appropriate post harvest chemical treatment to control superficial scald.
  (b) Slight risk of low temperature injury in fruit stored beyond May, especially in fruit with phosphorus concentration below 9 mg/100 g.

2 (a) Core flush risk after cool seasons or after prolonged storage in 5 per cent CO_2 regimes. Use 2 per cent O_2 level for storage beyond February.
  (b) For this regime ensure O_2 level does not fall below 1·6 per cent; carry out regular independent checks of O_2 levels using portable analyser and drawing atmosphere sample direct from store.

3 Liable to shrivel. This may be reduced by loosely covering the top layer of fruit in each bin with a polythene liner.

Table 2 Tentative storage recommendations for varieties of apple which are less widely grown or for which information is still being obtained

Variety	Air storage		Controlled atmosphere (CA)							Notes
				No scrubber			Using a scrubber			
	Temp °C	Terminate	Temp. °C	% CO_2	Terminate		% CO_2	% O_2	Terminate	
Charles Ross	3·0–3·5	November	3·5–4·0	5	December		5	3	December	
Chiver's Delight	4·5–5·0	December	—	—	—		—	—	—	1
Edward VII	3·0–3·5	January	3·5–4·0	8	May		5	—	January	1
Ellison's Orange	4·0–4·5	December	4·0–4·5	—	—		<1	3	January	2
Gala	0–1·5	February	3·5–4·0	—	—		<1	2	May	
Golden Delicious	1·5–2·0	January	1·5–3·5	8	April		5	2	April	3
Grenadier	0–1·0	Mid Nov.	—	—	—		—	—	—	
Holstein Cox	3·0–3·5	Mid Dec.	3·5–4·0	5	January		<1	2	Late March	
Howgate Wonder	3·0–4·0	January	3·5–4·0	8–10	April		—	—	—	
Idared	3·5–4·5	March	3·5–4·5	8	May		—	—	—	4
Ingrid Marie	1·5–3·5	*	1·5–3·5	6–8	*		<1	3	*	5
James Grieve	3·0–3·5	October	3·5–4·0	7–8	November		6	5	November	
Jonathan	3·0–3·5	January	4·0–4·5	—	—		6	3	November	
Kent	3·0–3·5	March	3·5–4·0	—	—		5	3	January	
Kidd's Orange Red	3·0–3·5	*	3·5–4·0	8	*		5	2	May	5
Laxton's Fortune	3·0–3·5	November	3·5–4·0	7–8	December		<1	3	May	
Lord Derby	3·0–4·0	December	3·5–4·0	8–10	March		5	3	Mid Jan.	
Lord Lambourne	1·5–2·0	January	3·5–4·0	8	May		—	—	Mid Jan.	
McIntosh	3·5–4·0	Mid Dec.	3·5–4·0	—	—		<1	2	Mid March	
Red Delicious	0–1·0	January	0–1·0	6–8	March		5	3	March	6

Variety									Notes
Spartan	1·0-2·0	January	1·5-2·0	—	—	<1	2	April	7
Sunset	1·5-2·0	January	3·0-3·5	8	May	—	3	—	
Suntan	3·0-3·5	February	3·5-4·0	—	—	5	3	May	8
Tydeman's Late Orange	3·0-3·5	February	3·5-4·0	5	March	5	2	April	
Winston	0·0-3·5	April	1·5-3·5	6-8	May	—	—	April	

Notes

1 Use post harvest chemical treatment to control superficial scald.

2 Susceptible to soft scald (a form of low temperature injury).

3 Risk of core flush and breakdown in some years at temperatures below 1·5°C in air or 5 per cent CO_2, 3 per cent O_2 and below 3·5°C in 8 per cent CO_2.

4 Risk of core flush in 8 per cent CO_2.

5 The approximate storage periods * for Ingrid Marie and Kidd's Orange Red have not yet been determined.

6 Pick mid-September.

7 Apply routine calcium sprays and/or post harvest calcium to prevent breakdown.

8 Risk of brown heart in 5 per cent CO_2 + 3 per cent O_2.

Table 3 Recommended storage conditions for the main varieties of pear (October 1977, by courtesy of East Malling Research Station)

(The terminal date for each set of conditions allows for a further 2 weeks at ambient temperatures to cover grading and distribution)

Variety	Air Storage		Controlled atmosphere (CA)				Notes
	Temperature	Terminate	Temperature	Using a scrubber			
				% CO_2	% O_2	Terminate	
Buerre Hardy	−1·0°C to −0·5°C	December	−1°C to −0·5°C	<1	2	Late April	1–5
Conference		March					1–6
Doyenne du Comice		March					1–5
Williams bon Chrétien		November					1–5

Notes

1 Picking dates very important for pears, especially when storing in CA. Use starch iodine test to judge maturity.

2 Set a second (overriding) thermostat to cut the refrigeration plant out when temperature of air leaving cooler falls to −2°C.

3 Ensure field heat is removed rapidly. Aim to cool to 4°C within 2–3 days of loading and obtain recommended storage temperature within a further 7–10 days. If fruit cannot be maintained below 0°C terminate storage earlier.

4 The minimum storage temperature should be raised by 0·5°C when soluble solids content of fruit are below 11 per cent, e.g. after cool, sunless summers.

5 Pears should be conditioned at a higher temperature after storage and before distribution and sale.

6 (a) In CA, ensure that CO_2 level remains **below** 1 per cent and O_2 does not fall below 1·8 per cent.
  (b) Cool to −1°C before establishing low O_2 regime.
  (c) Unscrubbed CA storage is no longer recommended for pears.

Marked differences in storage behaviour of fruit may occur from year to year in the same orchard. Differences may also occur between different trees and between fruits from different positions on the same tree in the same season.

Macroclimatic temperatures and total radiation cannot be significantly modified to the advantage of fruit production but the microclimatic conditions within orchards and within trees can be improved. Strong blossom, effective pollination and the rapid cell division and multiplication which occurs during the first three to four weeks after fertilisation of the ovules are all encouraged by warm weather. Though individually of marginal benefit the choice of sheltered sites, southern aspects and effective shelter belts can all contribute to better orchard environment. Orchard heating during the day to stimulate the activity of pollinating insects and pollen tube growth has also been tried with some success though further developments on these lines seems ruled out by present fuel costs.

The relationship between solar radiation and fruit size and colour has been amply demonstrated. Improved light penetration into trees can result from using particular tree shapes and suitable pruning techniques. Fruit size and structure can be markedly affected by irrigation. These factors are within the grower's power to control.

Seasonal weather conditions

The optimum keeping quality of apples and pears grown in the UK is generally reached in seasons of relatively high summer temperatures and radiation, and when there is no serious shortage of soil moisture, especially in the latter part of the season.

Low temperature injury of Bramley's Seedling and Cox's Orange Pippin is more likely following cool summers. It has been established that cold, wet weather during the four weeks prior to harvest increases the incidence of low temperature injury in Bramley's Seedling.

Examination of data from a ten-year study suggests a relationship between low mean summer temperatures and the incidence of core flush in Cox's Orange Pippin, though it is not yet known whether the higher humidity associated with dull rainy summers may also be implicated.

The incidence of rotting due to infection by *Gloeosporium sp.* is closely related to the total rainfall of August and September. Warm dry weather before harvest predisposes the fruit of certain varieties to suffer from scald. Bitter pit is also often more prevalent in dry seasons but is primarily the result of a number of interacting factors affecting the mineral composition of the fruit. Water core may be more prevalent after hot, dry summers but the mineral status of the fruit is also involved in determining susceptibility to water core.

Detailed descriptions of the above disorders and appropriate treatments to control or mitigate their effects on stored produce are given on page 28 onwards.

Permanent orchard factors

Relatively little is known about the effects of soil type on the storage quality of apples and pears. Interactions between rootstocks, availability of water and nutrition mask differences which may be due to soil type alone. Particular orchards have been noted for their consistent production of fruit of high storage quality and flavour but it has not, as yet, proved possible to identify the precise factors involved.

Bitter pit is generally more prevalent in fruit from orchards on sandy soils. There is evidence that potash is more readily taken up from sandy soils resulting in a higher ratio of leaf potash to magnesium and calcium than in the leaves of trees on loamy soils. But since water supplies tend to fluctuate in sandy soils this could unbalance the uptake of the different elements.

Evidence from overseas sources suggests a higher incidence of scald and internal carbon dioxide (CO_2) injury (brown heart) in apples from trees on light soils.

Investigations in progress with fruit from UK orchards may suggest correlations between certain soil types and the nutritional status and storage quality of the fruit, but at present the soil factor, other than its water holding capacity, seems of secondary importance to the effect of rootstock and the nutritional balance of the trees resulting from cultural practices in orchard management.

The incidence of bitter pit appears to follow the general pattern of rootstock vigour. The disorder is less prevalent on dwarfing than on vigorous rootstocks, though even dwarfing rootstocks may be affected where the crop is light or excessively vigorous growth occurs as a result of nutritional imbalance. The relationship between preferential uptake of potassium and the incidence of bitter pit in fruit from trees on M7, M26 and MM106 compared with M9 has been demonstrated experimentally but recent evidence suggests that such effects are only important when orchard conditions are marginal for the uptake of calcium. It is generally considered that fruit from trees on such dwarfing rootstocks as M9 and M26, should be picked slightly earlier than fruit from trees on the more vigorous rootstocks such as MM106 and M2.

Apples harvested from young trees are frequently susceptible to bitter pit and breakdown. This is associated more with oversized fruit and unbalanced growth than juvenility of the trees. As the trees age and cropping becomes more regular and vegetative vigour declines, so there is an accompanying decline in these storage disorders. On the other hand the occurrence of storage rots resulting from *Gloeosporium* infection increases as trees become larger and provide more sites for wood infection.

Variable orchard factors

The storage life and quality of apples and pears is markedly affected by cultural practices, most of which are under the control of the grower. Although there are still areas of uncertainty the general pattern of response to cultural practices is clear. The beneficial effects on fruit storage quality of changing the system of soil management from cultivations to grassing down has been recognised for many years. Grass reduces excessive tree vigour and mowing helps to regulate supplies of water. The availability of nitrogen, which is frequently in excess in UK orchards, is reduced by the presence of a grass sward.

The majority of bearing apple and pear orchards are down to grass with an area at the base of the trees treated with herbicides. Depending on tree spacing the herbicide treated area may be confined to individual trees or form continuous strips alternating with grass in the alleyways. A few orchards are grown in soils kept bare of all vegetation by overall herbicide application.

Partial elimination of the grass sward, which may be as high as 50 per cent of the orchard area in intensive plantations, or complete elimination as in the case of overall herbicide treatment, introduces entirely new conditions to the root environment and raises questions to which, as yet, the answers are incomplete. The response to these conditions cannot be judged on the experience of orchards which were previously clean cultivated, for then surface root destruction, soil disturbance and partial weed cover influenced growth and cropping. The relative proportions of grass and herbicide treated area affect the growth of trees, but there is little evidence of any direct effect of herbicides on tree growth or fruit composition.

Where herbicides replace mechanical cultivations the development of surface roots is encouraged. Water uptake then occurs after light summer rains, and phosphorus and potassium, which are more abundant in the upper layers of the soil, may be taken up more readily than from soils under grass. Excavations and other studies have shown that the roots of young trees growing in herbicide treated strips tend to be concentrated within the treated area. The establishment of strips may lead to a marked increase in soil acidity within the treated area which could result in calcium deficiency in the trees and fruit. Fruits from orchards under overall herbicide management appear to be lower in phosphate than where grass is present either in the alleys or overall.

The recent introduction of trickle irrigation systems in orchards raises further questions concerning nutrition and water relationships and their effects on storage quality, but as yet there is little firm information available. Under UK conditions, where irrigation is used to supplement natural rainfall, the risks of ill effects on storage quality are less likely than where trees are totally dependent on added water. It is, however, possible that excessive irrigation could lead to mineral imbalance or deficiencies in the fruit, and in some seasons the increase in fruit size resulting from irrigation may increase the incidence of breakdown and bitter pit in storage unless measures are taken to supplement fruit calcium levels.

Generally speaking where irrigation is used to alleviate stress its effects should be wholly beneficial, but if it is responsible for an imbalance in growth then there may be an increase in bitter pit and other storage disorders. More information is needed on the interactions between irrigation, nutrition and other orchard factors and the different soil management methods as they affect storage behaviour. Reduced rates of both nitrogen and potassium, and differential applications to grass and herbicide treated strips are currently being recommended, as is the application of calcium to the herbicide treated strips to correct incipient acidity.

Most of the investigations concerning the influence of applied fertilisers on the storage behaviour of apples and pears have been confined to the effects of nitrogen, phosphorus and potassium. More recently the effects of foliar and soil applications of calcium, and the interaction between calcium and potassium have been extensively investigated. The relationship between calcium and the incidence of bitter pit is discussed in the chapter Diseases and Disorders of Fruit, pages 29–31.

In Bramley's Seedling the incidence of low temperature breakdown, (LTB) which is normally associated with storage at temperatures below 3·5°C has also been shown to be partly related to low levels of calcium and phosphorus in the fruit. Excessive potassium and a high potassium to calcium ratio in the fruit is also associated with an increase in LTB in this variety. Levels of these elements in leaves and fruit can be determined as a guide to the likely storage behaviour of the fruit.

As evidence accumulates on the influence of nutrition on fruit storage quality, the need for greater precision in the nutritional management of the trees as affected by fertiliser practice becomes increasingly apparent.

Three aids to deciding the appropriate nutritional policy are soil, leaf and fruit analysis.

Soil analysis indicates the amount of available nutrients present in the soil. Analytical results from ADAS are expressed in terms of units of fertiliser according to indices for the levels of available phosphorus (P), potassium (K) and magnesium (Mg.). (see p. 11).

Interpretation of *soil analysis* data:

1. When a K index of 0 is encountered before planting a large dressing of potash (up to 350 kg/ha K_2O) should be ploughed down, in addition to the normal application recommended.

2. When the K index for the 0–15 cm depth sample is less than 2 and that of the 15–30 cm depth sample is 0 the potash recommendations should be increased by 60 kg/ha K_2O.

3. Muriate of potash (60 per cent K_2O) is suitable for tree fruits except in coastal areas where soil chlorides are likely to be high and sulphate of potash should be used.

4. At P indices greater than 0, K indices greater than 1 and at all Mg indices, twice the annual rates can be applied every other year.

5. Nitrogen (N) recommendations are intended as a guide and should be varied according to variety, rootstock vigour, appearance of foliage and use of irrigation. Soil characters affecting depth of rooting and moisture reserves will influence the nitrogen requirements. More nitrogen (50 kg/ha) will be required on coarse textured soils, shallow soils and on soils with a poor structure.

6. Mulches. Where FYM is used the rates of application of nutrients should be reduced (for every 10 tonnes the nitrogen should be reduced by 15 kg; phosphate by 20 kg; potash by 40 kg and magnesium by 8 kg). Where straw is used, the rates of application of potash should be reduced by 8 kg/ha for every tonne of straw.

Leaf analysis indicates the current nutrient status of the tree and shows how effectively it is using the nutrients which are present in the soil. Leaf analysis can be used for assessing the nitrogen requirements of the tree, but is less reliable than mineral analysis of the fruits as a guide to the storage behaviour of apples. Where possible data from both sources should be used.

Table 4 Fertiliser recommendations (kg/ha annum) for top fruit in areas where summer (April–September) rainfall is less than 350 mm

Fertiliser/Crop	N Cultivated	N Grass/Herbicide Strip	N Grass	P Index 0	1	2	3	3+	K Index 0	1	2	2+	Mg Index 0	1	2	2+
	N			P_2O_5					K_2O				Mg			
Apples—dessert	40	60	120	80	40	20	20	Nil	220	160	80	Nil	60	40	30	Nil
—culinary	100	120	180	80	40	20	20	Nil	220	160	80	Nil	60	40	30	Nil
Cherries Pears Plums	90	120	150	80	40	20	Nil	Nil	220	160	80	Nil	60	40	30	Nil

Leaf samples should be taken from the middle of the current season's extension growth in mid-August. Leaves taken at other stages of growth or from different positions show marked differences, particularly in calcium content. For mid-August, middle position leaf, the following standards are suggested: (Leaf nutrient levels expressed as per cent dry matter).

N	P	K	Mg
2·4–2·6	0·20–0·25	1·20–1·55	0·20–0·25

RELATIONSHIP BETWEEN INDIVIDUAL LEAF NUTRIENTS AND STORAGE DISORDERS

Although fruit analyses are better than leaf analyses as a guide to storage behaviour, the following relationships between leaf nutrients and disorders are important.

Nitrogen

There is very little experimental data available on the optimum levels of leaf N. However, susceptibility to rotting, loss of firmness and poor colour are all associated with increasing levels of leaf N and, since there is apparently little yield advantage in increasing leaf N above about 2·6 per cent, this figure is taken as a reasonable maximum. In practice, leaf N levels are generally well above this value in English orchards and there seems to be a general need to moderate dressings of nitrogen in orchards of Cox's Orange Pippin.

Phosphorus

There is increasing evidence that where leaf P levels fall below about 0·24 per cent, there is a tendency for fruit to soften prematurely in store or even develop breakdown. This is aggravated where calcium (Ca) levels are also low.

Potassium

Leaf K levels are a useful guide to the risk of bitter pit. A maximum leaf K figure of about 1·55 per cent is recommended but this must be considered in relation to both the Ca levels and fruit size at harvest.

Magnesium

There is very little information available on the relationship between leaf Mg and storage quality. Magnesium is inversely related to K in the leaf but positively related to K in the fruit. Hence, the information obtained for fruit Mg cannot be directly related to the leaf levels.

Calcium

In most seasons leaf Ca levels provide a general guide to the risk of bitter pit and breakdown, particularly when considered in relation to leaf K and leaf P, but fruit size at harvest will markedly influence the interpretation of the data. The amount of rainfall between the sampling date in August and harvesting can also influence the relationship between leaf Ca and the subsequent risk of bitter pit in the fruit. Hence, for bitter pit prediction purposes in all seasons a sample of the fruit at harvest is more reliable than an August leaf sample.

RELATIONSHIP BETWEEN INDIVIDUAL FRUIT NUTRIENTS AND STORAGE DISORDERS

The relationship between mineral analysis of fruit can be used to diagnose and predict the causes of storage disorders. Analyses of apples from store may indicate if mineral imbalance is involved in the occurrence of any particular disorder. However, from fruit analysis alone, it is not always easy to make changes in orchard nutrition which would help to reduce disorders in future seasons. This is because fruit composition is affected by the interaction between nutrients, season, soil conditions and orchard factors, particularly those affecting fruit size and moisture relationships.

To predict the normal commercial harvest date of Cox's Orange Pippin, the following standard is suggested:

N	50–70
P	11
K	130–160 (mg/100 g fresh weight)
Mg	5
Ca	4·5

Apples of this composition should keep satisfactorily for the periods stated in the recommendation for this variety in Table 1.

Nitrogen

The higher the N content the more the fruit is susceptible to rotting, loss of firmness, poor skin finish and poor colour. Above 80 mg/100 g the risks of the above disorders occurring are high and below 60 mg/100 g the risks are low.

Phosphorus

As fruit P falls below 14 mg/100 g there is an increased risk of the fruit losing its firmness and developing breakdown. This is worse where P levels are below 10 mg/100 g and where calcium levels are also low (see Table 6).

Potassium

The higher the level of K in fruit the higher the risk of bitter pit. The incidence of the disorder is greater the lower the Ca/K ratio (Table 5). There is evidence that as the level of K increases so does susceptibility to *Gloeosporium* rotting and core flush, but whilst the initial levels for rotting appear to be similar to those obtained for bitter pit, it is premature to suggest any standards for the latter. Generally speaking fruit flavour and acidity increase with increasing levels of K. Thus if Ca levels are adequate (over 5·5 mg/100 g), high levels of K can be advantageous in terms of fruit quality.

Calcium

Fruit Ca levels, when considered in relation to K and P, are a good guide to the risks of bitter pit and breakdown respectively. Fruits are also more susceptible to *Gloeosporium* at low Ca levels.

Disorders of fruit in store are also discussed in the special chapter, pages 28–34.

Table 5 Tentative standards for predicting the incidence of bitter pit and *Gloeosporium**

Ca (mg/100 g fresh weight)	K (mg/100 g fresh weight)		
	170	130–170	130
5·9	Very slight	Very slight	Very slight
5·5–5·9	Slight	Very slight	Very slight
5·0–5·4	Medium	Slight	Very slight
4·5–4·9	High	Medium	Slight
4·0–4·4	High	High	Medium
4·0	High	High	High

The risk of *Gloeosporium is dependent on the level of inoculum in the orchard and is influenced by fruit skin finish, fungicide programme and the amount of rainfall during the four weeks preceding harvest.

Table 6 Tentative standards for predicting the incidence of breakdown†

Ca (mg/100 g fresh weight)	P (mg/100 g fresh weight)		
	14	10–14	10
5·4	Very slight	Very slight	Slight
5·0–5·4	Very slight	Slight	Medium
4·5–4·9	Slight	Medium	High
4·0–4·4	Medium	High	High
4·0	High	High	High

†Senescent breakdown and loss of firmness are both adversely influenced by late picking. Although susceptibility to low temperature injury is also partly related to low fruit Ca and P levels, this disorder is uncommon in Cox's Orange Pippin stored in air at 3°C.

EFFECTS OF SPRAY MATERIALS

The skin finish of apple and pear fruits may be affected by spray materials applied for the control of pests and diseases.

The influence new materials may have on the composition of the internal atmosphere of the fruit may remain largely unrecognised because, in many cases, trials with new materials are not extended into storage.

However, limited information suggests there is a possibility of side effects which might influence storage quality.

Benomyl applied either pre-or post-harvest has been shown to have some influence on ripening and the incidence of a number of storage disorders, but further work is required to determine the extent and conditions under which side effects may arise. Dimethyl, dithiocarbamate sprays applied for the control of apple scab have been reported to lead to an increase in bitter pit in some varieties, possibly through reducing the transpiration rate of leaves and fruit. Dutch research workers have shown that the incidence of scald in Golden Reinette apples was higher where the fruit had received sprays containing organic mercury than where sulphur had been applied. Thiram also gave a very marked increase in scald in this variety, but since fruits which develop scald are also frequently affected by bitter pit it is possible that the increase associated with thiram was also due to its effect on the mineral composition of the fruit.

The replacement of lime sulphur by organic fungicides for the control of apple scab has been coincident with an increase in the problem of bitter pit. But recent studies indicate that Ca supplied in foliar sprays at the same rate as that in a lime sulphur programme, does not help in the control of bitter pit in fruits of Cox's Orange Pippin and Worcester Pearmain. The improved leaf and skin finish resulting from the replacement of lime sulphur by organic fungicides evidently has a profound effect on fruit growth, size and composition and this has increased the susceptibility to the disorder.

Ethrel is used to promote ripening and fruit colouring by supplying the ripening agent ethylene in advance of its natural development. There is little evidence of any effect on storage behaviour of apples when the picking date is adjusted to take account of the earlier onset of the climacteric. 2,4,5-TP may be used in conjunction with ethrel to delay fruit abscission while the fruit colour develops. When used within two weeks of harvest there have been no significant effects on the storage behaviour of Cox's Orange Pippin apples.

There is less evidence on the effect of these chemicals on pears, although premature breakdown of William's Bon Chrétien during ripening after storage for 50 days at 0°C has been reported.

TIME OF HARVESTING

Ideally the optimum time of picking is just before the climacteric, that point in ripening when a rapid rise in the respiration rate occurs. Determination of the optimum time for harvesting fruit which is to be stored is not easy. Many attempts have been made to correlate optimum picking date with some point in the growing cycle of the tree. For the variety Cox's Orange Pippin, the number of days between petal fall and onset of the climacteric (that is the climacteric of apples, picked not more than two days earlier and held at 12°C) has been recorded: in experiments over nine years the number of days ranged from 120–144. Yet in those nine years, the optimum harvest date lay between 20 and 24 September. (But see also page 8.) Although not all fruits can be picked on the same day, foreknowledge of the time of the climacteric provides a date around which the work must be organised if the storage disorders associated with picking too early or too late are to be avoided. Present evidence suggests that satisfactory long-term storage of Cox's Orange Pippin can be achieved by picking up to a week before or after the onset of the climacteric.

The picking date of other late storing varieties can be related to that of Cox's Orange Pippin and the optimum date for this variety under Kentish conditions can be obtained each season from ADAS. The optimum dates for other areas can be adjusted accordingly but experience of local conditions over a period of years will remain the best guide.

Successful storage of pears is critically dependent on determining the optimum picking date. Probably the best method is the starch/iodine test now widely used in commercial practice. With this method it is possible to follow the reduction in starch content as ripening progresses: it is described in the *East Malling Research Station Annual Report* for 1970 (1971) pages 149–151, from which, with permission the following section has been taken. Most of the experiments were with the variety Conference to which the following data primarily relate but the method is equally applicable to other varieties though the pattern of staining may be different.

The use of the starch/iodine staining test for assessing the picking date for pears

The starch/iodine reaction forms the basis of a fairly simple test for judging the maturity of pears.

Starch accumulates in the pears during growth, and in a fine summer the cells become full of starch during the later stages of growth. This may be shown by applying a solution of iodine to a cut surface; starch stains blue-black. The intensity of the stain is usually greatest near the skin and in the core region. Some ten days or so before climacteric, starch starts to disappear and the area which stains with iodine, as well as the intensity of the stain, decreases.

To carry out an iodine test a sharp knife is needed to cut pears in half equatorially and a one per cent solution of iodine in four per cent potassium iodide (available from local chemists' shops). The solution of iodine is applied to a freshly cut surface and the pattern of staining may be noted immediately, but a more reliable assessment can be made about an hour later. In order to discover when starch begins to break down, it is necessary to note the patterns of several fruits, two or three times a week, beginning two or three weeks before the usual picking date. It is best not to pick while the cut fruit stains uniformly dark blue over the whole surface. When the blue-black pattern has decreased by about one-third of its maximum coverage, the fruit should be harvested. It is important to note that the fruit should be tested immediately it is picked because the conversion of starch to sugar occurs quite rapidly in picked fruit. It may also be noticed that in some varieties the pattern is not regular and takes on a mosaic effect over the whole surface of the cut section but whatever the pattern the same assessment of the decrease in the area of the starch pattern applies.

Typical starch patterns for Conference pears, at the East Malling Research Station in 1970, are shown in Figs. 1 and 2. The sections shown in Fig. 1 have been selected from a range of samples taken every two to four days, and show the major features to be noticed. The dates of picking were: (A) 24 August; (B) 1 September; (C) 7 September; (D) 10 September; (E) 14 September; and (F) 24 September.

16

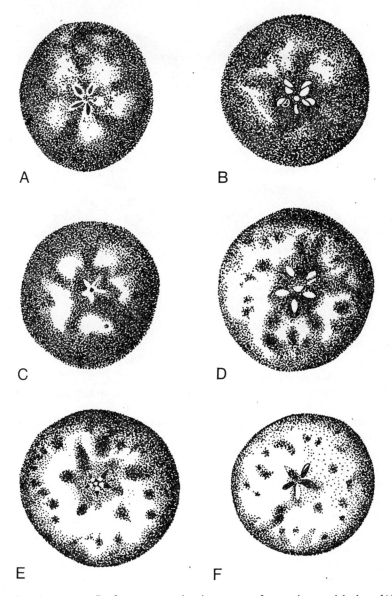

Fig. 1   Starch patterns, Conference pear, showing average for specimens picked on (A) 24 August; (B) 1 September; (C) 7 September; (D) 10 September; (E) 14 September; (F) 24 September (re-drawn from Photo: E. Malling Res. Stn.)

The four sections shown in Fig. 2 were taken on 7 September and illustrate the range of starch patterns seen at any one date. It is not too difficult to judge the average amount of starch; for example in Fig. 1 (A) it is 75 to 80 per cent and in Fig. 1 (B) 90 per cent. The amount of starch had fallen to somewhere near two-thirds of the maximum (shown in Fig. 1 (B)) by a date midway between those on which Figs. 1 (C) and 1 (D) were photographed, that is about 8 to 9 September, and this was the right date on which to start picking these particular pears.

Pears may be stored in controlled atmosphere, or in air. The CA conditions

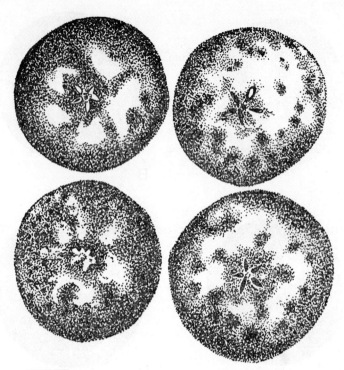

Fig. 2   Starch patterns, Conference pear, showing range of intensity obtained in specimens picked on the same date (7 September 1970) (re-drawn from Photo: E. Malling Res. Stn.)

for Conference pears were formerly 5 per cent CO_2, 5 per cent O_2, at 0·5–1·0°C but less than 1 per cent CO_2, 2 per cent O_2, at −1°C is now preferred for long term storage. If the pears are too mature when picked they are injured by the CA conditions. The starch/iodine test enables a decision to be made as to whether CA storage will be safe or not. This safe period usually lasts some five to six days after the optimum date for the start of harvest, and Fig. 1 (E) shows the starch pattern which was judged to be the minimum tolerable in 1970. The optimum date was 8 to 9 September, and the end of the safe period for CA was thus 14 September.

For storage in air it is recommended that the air coming off the cooler should be at −1·5°C; this gives an average pear temperature of 1·0 to −0·5°C. The optimum picking stage for CA (Fig. 1 (C–D) can be used for storage in air, but whereas the harvesting period for CA was limited to some five to six days, it can be considerably extended for storage in air, up to or even beyond that which gives the type of starch pattern shown in Fig. 1 (F).

Pears continue to increase in weight up to the time they are picked. The average weight of those picked on 8 to 9 September was 160 g; on 14 September (end of period of suitability for CA storage) it was 170 g, and on 24 September, 190 g. Thus the weight increased by 11 per cent between 14 and 24 September.

This test is less reliable for judging maturity of apples. The best time to harvest apples is just before the climacteric, as discussed earlier.

Picking apples and pears at times widely divergent from the optimum date results in a number of storage disorders. Early picking of apples of susceptible varieties may result in superficial scald or core flush. External carbon dioxide injury seems to occur mainly on early picked apples but it can also be found on fruits picked at the correct time and stored in a cold spot in normal CA conditions. Superficial scald does not normally occur on English pears unless they are over-ripe. Exceptions are Packham's Triumph and Anjou on which varieties scald may appear during storage. Early picked fruit loses water and shrivels more readily than fruit picked at the optimum stage and being smaller than fruits harvested later, is an unnecessary loss in yield. Pears especially are able to increase their size significantly during the last two or three weeks before harvest but if left too late are more prone to internal breakdown.

CONTAINERS AND THE ORGANISATION OF PICKING

It is a vital necessity when picking and loading fruit into store that the operations should be done at the optimum time and that the mechanical damage to the fruit should be kept to the absolute minimum.

To obtain the best results when storing *Cox's Orange Pippin, seven days before or after the optimum picking date is the maximum that should be allowed.* Other varieties may have greater or lesser tolerances. It follows that the harvesting of a large area of any one variety is a major exercise in planning and organisation. Accurate estimation of the yield is crucial to planning the number of pickers, and the number of containers required and their positioning in the orchard.

The equipment needed for conveying fruit from the orchard to the store or packhouse depends on the total weight of crop, the type of container, the lines of communication, and the storage capacity. The cheapest method in terms of capital investment is a tractor and low loading trailer on which the full boxes are loaded by hand and unloaded at the packhouse (Fig. 3 gives comparative data for various systems). However, the paramount importance of timeliness in harvesting and storing, together with a reduction in costs achieved with unit loads, has led to the use of pallet systems for box handling. From these has developed the extensive use of bulk bins for apples and on a more limited scale for pears. Bins have also led to a reduction both in the amount of mechanical damage caused to the fruit and in the time taken for harvesting.

Overall labour using bins is estimated at 54 per cent less than that using loose bushel boxes and 24 per cent less than that using palletised boxes. Bulk bins should be manufactured with dimensions of the International Standard pallet of 1000×1200 mm and with a depth c. 500 mm. This will give a capacity of approximately 270 kg depending on the variety of fruit. Storage in bins uses about 30 per cent less timber than boxes; thus the quantity of water absorbed by the wood from the fruit and shrivelling are reduced. Correspondingly a greater weight of fruit can be held in a given store space when bins are used though it is important to ensure that the refrigeration plant has sufficient capacity to cope with the increased rate of loading and total quantity held in store.

Bins may be well-based or pallet-based. Well-based bins allow a greater density of storage and are more robust and rigid. They are also easier to man-handle on the ground and easier and quicker to pick up, while they travel more steadily on tractor forks than the pallet-based types. Their disadvantages are less manoeuvreability in the store because of the need to lift them 'square',

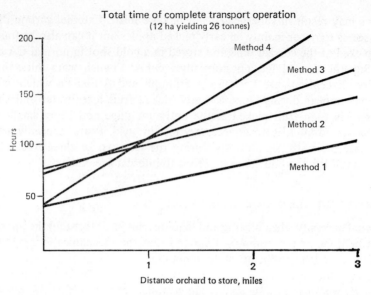

Fig. 3a Transport of apples from orchard to store or packhouse: comparative data of four methods

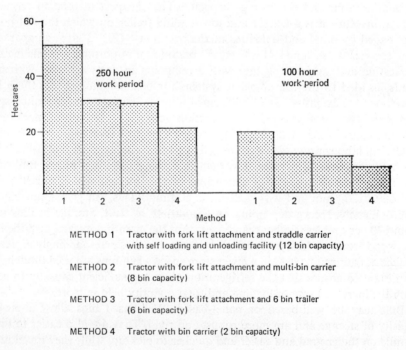

METHOD 1   Tractor with fork lift attachment and straddle carrier
            with self loading and unloading facility (12 bin capacity)

METHOD 2   Tractor with fork lift attachment and multi-bin carrier
            (8 bin capacity)

METHOD 3   Tractor with fork lift attachment and 6 bin trailer
            (6 bin capacity)

METHOD 4   Tractor with bin carrier (2 bin capacity)

Fig. 3b The effect of handling method and duration of harvest period upon area cleared using one set of equipment, carting to store 1 mile (1·6 km) distant whereas pallet-based bins can be lifted on forks withdrawn on the skew when working in confined spaces. There is also a need for a second set of forks if pallets are used for bulk handling elsewhere on the farm.

Bulk bins may be constructed of plywood or planking. Plywood is preferable as less damage to the fruit is caused but it is important that bins for pears be treated with a water repellant compound to prevent the diffusion of harmful chemicals from the surface of the timber. Skin staining of pears in contact with the sides of plywood bins has been found to occur when the fruit and bins were damp.

Bulk bins represent a considerable capital investment. Storage in the open will cause splitting and buckling and lead to premature deterioration. Bowing of the sides may occur with consequent difficulties in store loading. Ideally bins should be stored under cover in empty stores or barns or sheeted down under black polythene. Bins of dimensions 1000 × 1200 mm can be nested. They should be marked with the year of manufacture and numbered for easy identification.

PRE-STORAGE TREATMENT AND HANDLING INTO STORE

All unit load handling around the buildings can be done with fork-lift trucks. It is essential that all floors are on the same level.

Pre-storage treatment of apples with antioxidants and/or fungicides by dipping or drenching has become a routine operation in many organisations. Details of the treatments are given on page 29. There is a range of proprietary equipment available for handling fruit and growers have made their own installations to meet particular needs.

After picking and any pre-storage chemical treatment it is essential that the fruit be placed in store and cooled to the recommended temperature as quickly as possible. The store size should be such that it can be loaded with freshly picked fruit in four days and the refrigeration plant should be capable of reducing the temperature of the fruit to 4·4°C each day so that the final storage temperature is reached three days after completion of loading.

Apples picked in hot weather may be left in the open overnight to rapidly lose heat and loaded into store early the following morning. Apples in bulk bins stood out overnight can lose up to 70 per cent of their heat through radiation during the hours of darkness. At no time should fruit be allowed to stand near anything that is likely to cause taint. Fresh paint, for example, quickly ruins the flavour of apples and pears.

Time and motion studies have shown large variations in the time taken to stack bulk bins in the stores. Only the most experienced fork-lift truck driver can guarantee to place the lower bins accurately enough to ensure correct stacking of the upper bins. This is important when there is little spare space in the store. With age the sides of the bins may begin to bow, increasing their width by as much as 50 mm and adding 250 mm to a six-wide stack of bins. This may cause difficulties in completely filling the store and would in any case vitiate the effect of the 20–30 mm gap between each stack, left to ensure air circulation. When designing new stores sufficient room should be left for errors in bin placement and the bowing of bins with age. With existing stores much depends on the skill of the loader but he can be helped by a wooden template which accurately positions the wooden blocks and bottom bins. Painting white lines on the store floor is another, but less flexible, possibility.

Ideally, stores should be loaded to capacity. Satisfactory results cannot be expected from stores which are only partially filled. Twice daily readings should be taken of the temperatures and gas concentration in the store; these together with opening and closing times should be logged. A close check on the atmospheric conditions can thus be kept and will provide valuable information if the fruit, on being taken from the store, is not in a satisfactory condition.

Soft fruits

Strawberries, raspberries, black currants and other bush and cane fruits cannot be stored for any length of time unless frozen. Deterioration due to natural senescence, together with the effects of handling, sets in soon after removal of the fruit from the plant. The most that can be done is to reduce loss of quality by careful picking, packing, handling and transport together with cooling at a stage appropriate to the method of marketing and the crop concerned. Cooling is not normally necessary for fruit for processing which is either picked under-ripe, pulped or juiced. However the development of nitrogen freezing plants for the preservation of whole raspberries demands a supply of top quality fruit which may benefit from pre-cooling. This is a limited development at present but could extend to other fruits destined for eventual sale in wholesale or retail freezer packs. The care required in handling fruit for these outlets will be similar to that required for the fresh fruit market.

The object of cooling is to remove the field heat, that is heat resulting from solar radiation and high ambient temperatures, and the metabolic heat generated by the respiration of the fruit. The growth of moulds is accelerated by high temperatures and humid conditions and occurs more readily on over-ripe and damaged fruit. It also adds further metabolic heat to the fruit. For these reasons it is important to pick at the correct stage of ripeness, to avoid bruising, tearing and other mechanical damage and to reduce or limit the rise in fruit temperature. Respiration rates for different kinds of fruit are given in Appendix 4 (b).

STRAWBERRIES

Different methods of marketing will affect the need for, and extent of, cooling and may influence the choice of container for the fruit. Fruit is marketed direct to local outlets or to chain stores or distributed through central markets. In the first case pre-cooling is not usually required, though the quality of the fruit will be enhanced and its shelf life prolonged if it is prevented from overheating in the field by being kept in the shade after picking. Cooling may be a desirable preliminary to transport to distant centres or an essential part of the marketing process in the case of cool chain marketing.

It has been amply demonstrated that the storage life of strawberries is increased as temperature is reduced. It is eventually limited by rotting of the fruit and loss of flavour. Although under ideal conditions fruit picked early in the season has been stored for three weeks at 0°C, in commercial practice *only storage for periods of up to five days at 2–3°C has proved satisfactory*. Good storage for this length of time is dependent on the fruit being disease free and undamaged. A full spraying programme against *Botrytis* is essential to limit the infection with spores which, even if prevented from developing during storage, may cause problems as the fruit re-heats in the market or retail shop. Fruit picked later in the season is likely to carry many more spores than earlier picked fruit and therefore tends to rot more quickly from infection by *Botrytis*, *Mucor* and *Rhizopus spp*.

Picking and packing

Fruit must be graded as it is picked and placed directly into the consumer unit for retail sales, normally 250 g and 500 g punnets. The design of punnets is important, they must permit ventilation and adequately protect the fruit; they ought to be big enough to hold the required quantity of fruit below the rim of the punnet. Their fabric varies: moulded paper punnets are absorbent and have gentle corners but in the moulding process, the edges of the ventilation slots are turned inwards and can cause damage to the fruit. Waxed cardboard punnets have no sharp internal edges but are less robust and the corners and top edges are right-angled and sharp. Plastic punnets are flimsy but when well designed and without sharp edges, are satisfactory if contained in an outer tray. Outer containers vary greatly and may be constructed in fibreboard, wood or plastic. Closed and closely constructed containers offer maximum impedance to the air flow, and thus to heat extraction, especially when closely stacked. The corner-posted stacking tray, manufactured in wood is one of the most satisfactory containers combining good ventilation with rigidity. One tonne of fruit will occupy about 8 m³ when packed in these wooden trays (allowing a 20 per cent margin for pallets and spaces between stacks).

Since ripening and the development of fungal growth on the fruit are both aggravated by damage it is of paramount importance to pick only sound fruit at the correct stage of ripeness, that is when just under fully ripe. Individual fruits must be handled once only and transported from the field with the greatest of care and the minimum of movement. Pallet loads handled by fork lifts on the tractor are to be preferred to moving each tray by hand. Sprung trailers with canvas sun shields and equipped with low pressure tyres will minimise the risk of damage to the fruit.

Cooling and transport for marketing to wholesale and retail outlets

Where fruit is to leave the farm on the evening of the day it is picked there is no advantage in cooling it below about 10°C. The fruit will warm up quickly after it has left the cold store and will continue to ripen and deteriorate at the same rate that it would have done before the cooling operation. Cooling to 10°C can be achieved in conventional apple stores, though it may be necessary to partition the chamber in order to increase the rate of air flow around the fruit (see following section, Cool chain marketing).

After removal from store, fruit should be transported to market as soon as possible. In most cases this will be by lorry. In order to delay the increase in

23

temperature and protect the fruit from wind damage, dust and fumes, loads should be sheeted down. The fruit of an unsheeted lorry load, transported during the day, is likely to reach ambient temperature within half an hour of leaving the farm. Unsheeted loads transported during the night will normally lose heat, but the effects of physical damage from wind may be severe and the best practice is to pre-cool the fruit and transport it in covered vehicles.

A recent development in improving the storage life and market appearance of strawberries is the use of the *Transfresh* system. In this system refrigerated, unit loads of fruit, in sealed trucks or sealed packages, are subjected to modification of the atmosphere by the introduction of various gases within the load. The system of sealing the units and composition of the gases are subjects of patent rights owned by the Whirlpool Corporation of America for which American Fruit Importers Ltd of Western International Market hold the UK agency. Experience over several seasons has demonstrated to the satisfaction of users that treated fruit has a much enhanced life and improved condition compared with untreated fruit. This has been reflected in consistently better market prices which have more than covered the cost of the treatment. Treated fruit must be of high initial quality since the system can do nothing to improve the quality of poor fruit.

Cool chain marketing

This method of marketing aims to exploit the benefits of cooling the fruit after harvest, and the preservation of quality by keeping the fruit at a temperature that will delay deterioration throughout the distribution period. Its success depends on a contractual arrangement between grower and retailer so as to co-ordinate the supply of fruit with demand and, on the provision of refrigerated transport to maintain the cool chain from farm to retail counter.

Where cold chain marketing is involved it is essential to pick only top quality fruit. The benefits from cooling are somewhat speculative and it is difficult to decide what level of capital investment is justified, but any advantage will be entirely vitiated if the fruit is not of superlative quality at the start.

Since there are considerable difficulties in co-ordinating the rate of supply and demand, it is vital to the success of the system that the production site should have equipment for both rapid cooling of the fruit and temporary storage at a suitably low temperature. Evidence to date suggests that the temperature of fruit should be reduced to about 4°C within six hours of picking and then held at a temperature of between 2 and 4°C until required for delivery.

The equipment necessary to obtain these objectives depends on the volume of fruit to be handled. A modified apple store may be adequate for the purpose, especially if fruit can be loaded within an hour or so of picking, but in other cases equipment purposely designed for rapid cooling may be necessary. Investigations on methods of rapid cooling are being made by the ARC Food Research Institute at Norwich, while research on vacuum cooling of unit loads is being undertaken at the National Institute of Agricultural Engineering, Silsoe. There is also commercial interest in a system of cooling which involves blowing refrigerated air into a complete marketing unit. Advice on the current stage of these developments can be obtained from ADAS mechanisation advisers.

24

The refrigeration capacity of controlled atmosphere fruit stores is usually 50 W/m³. During the summer it can be assumed that approximately half this will be available for cooling the fruit while the remainder is utilised in maintaining the store temperature. Assuming a cooling capacity of 34 W/m³ and that the fruit will be cooled to 4°C the maximum loading rates for different ambient conditions can be determined from the following table. It should be noted that fruit temperature does not necessarily equal air temperature and in hot weather it is highly desirable that fruit awaiting loading into store should be shaded from the direct rays of the sun under well ventilated cover.

Table 7   Maximum loading rates for soft fruit at two different air temperatures and two cooling periods (kg/cm³ h)

Ambient Conditions	Air Temperature	3 h cooling	4 h cooling
Hot	21°C	2·6	2·8
Warm	16	4·0	4·2
Cool	10	8·0	8·4

The loading capacity in kg/h for a store is found by multiplying the store volume (m³) by the appropriate factor from the table: for example a 170 m³ store; hot ambient conditions (21°C) and 3 h cooling has a maximum loading rate of 2·6 × 170 = 442 kg/h. In order to achieve the required cooling times the minimum air flow must be 0·0007 m³/s kg (0·0007 cubic metres per second per kilogram) and 0·0005 m³/s kg for the three and four hour cooling times respectively. Most fruit stores have a fan capacity of 0·01 m³/s m³ and are likely to circulate sufficient air to give the required cooling time, provided that it is concentrated on the fruit. When fruit is being loaded during the day it is likely that the air temperature will rise above the required 2°C but it should fall during the night and, providing the calculated loading rates are being observed, there should be no problems.

During the cooling process up to one per cent of the weight of the fruit may be lost and it may be greater if cooling takes appreciably longer than six hours see also pages 121–3.)

Covering punnets with paper or film may be a marketing requirement. If this is the practice for cool chain marketing, or with any fruit subjected to a period of pre-marketing cooling, and the punnets are wrapped before cooling, heat flow will be impeded. Condensation will occur whether the punnets are wrapped before or after cooling. The film cover acts as a barrier to air flow and prevents the transfer of moisture from the fruit to the cooling air. Fruit should preferably be cooled before it is over-wrapped and, when destined for sale from cool counters, be transported in refrigerated containers. The risk of condensation forming on over-wrapped punnets of fruit to be sold from non-refrigerated counters can be mitigated by transporting in closed insulated containers in which the fruit warms up gradually.

The matters discussed concerning strawberries apply with even greater emphasis to the handling, cooling and marketing of raspberries. Though strawberries are more liable to mechanical damage than raspberries, the results of damage to the latter are more immediately obvious because of settlement or compaction in the punnets and rapid invasion of the tissue by moulds and rots.

The bulk of Scottish fruit is sold to processors located in the area of production for canning, freezing or pulping. There is no tradition of long distance marketing for retail sales from any of the production centres in the UK although there have been many trial sendings. The development of varieties suitable for mechanical harvesting is not likely to alter this situation, except insofar as they would need to be more resistant to mechanical damage and thus more suitable for short-term storage and wider distribution from the point of production.

Other fruits

No detailed investigations have been made on the cooling and handling requirements of other soft fruits. Blackberries and loganberries may be expected to respond in a similar way to raspberries. The organisation of black currant harvesting for processing is such that no immediate post-harvesting treatment, except the obvious one of shading the fruit while awaiting collection, seems called for. Fruit for processing may be juiced, blast-frozen or stored in modified atmospheres containing high levels of CO_2 by the manufacturers in preparation for subsequent use.

Gooseberries, as relatively hard fruit, can be cold stored for a limited period. Most fruit is sold direct to processors for canning and freezing and the need for storage does not arise. Immature fruits of the dessert variety Leveller have been stored successfully for five days at 0°C but both flower and stalk end were susceptible to attack by *Botrytis sp.*

Plums may be stored for a short period before marketing. Investigations at Luddington Experimental Horticulture Station have shown that Victoria plums can be stored for three to four weeks at 1°C before they become unmarketable due to softening. The fruit should be picked just before full ripeness. Less ripe fruit tends to lose flavour and to develop a tough skin. Marjories Seedling can be stored for up to three weeks but after this deterioration is rapid.

Storage diseases and disorders of fruit

The spoilage of fruit during refrigerated storage can be caused by one or more of several different factors. It may be caused by fungal infections or by functional or physiological disorders.

The extent of loss from physiological (non-parasitic) disorders depends on such factors as soil conditions, weather conditions and cultural practices during the time the fruit is developing on the plant, on the degree of ripeness of the fruit at picking, on the temperature and atmospheric conditions maintained during storage, on the length of time the fruit is stored, and on the time interval between picking and storage.

The level of wastage associated with fungal attack may depend on the effectiveness of the pesticide spray programme, the degree of mechanical damage at harvesting, on any post-harvest, pre-storage pesticide treatments, on the temperature and atmospheric conditions of the store, and the length of time the fruit is stored.

Physiological disorders are sometimes difficult to distinguish from each other and the cause of some of them is open to conjecture. Some develop before the fruit is harvested and increase in severity during storage whilst others develop as a response to adverse storage conditions. The storage disorders and diseases of apples and pears have been most extensively studied, whereas information on storage disorders of other fruit crops is limited.

Storage disorders of apples

SUPERFICIAL SCALD

Symptoms

Symptoms of superficial scald (see Plate I) mostly occur on fruits of green skinned varieties but when coloured fruits are affected the damage is usually confined to the greener side of the fruit. Varieties which are especially prone to the disorder include Bramley's Seedling, Edward VII and Granny Smith.

Symptoms can vary depending on the variety, and different stages in the development of scald can occur within the same consignment of fruit. The stages of superficial scald can include 'bronzing' in which areas of skin have a bronze tint with the lenticels standing out as prominent islands of green. More severe superficial scald includes patches of skin which are light brown in colour and these may or may not be slightly sunken. In severe cases the patches are dark or light brown in colour and sunken and the underlying cortical tissues are damaged and the dead layers of tissue can slough away from the underlying healthy tissue.

Causes

Scald is usually more severe on early harvested fruit and on large apples. It is also more severe on fruit harvested after a warm dry summer; it is likely that this is an effect of water stress.

It has been shown in experiments that for Edward VII stored until May a close correlation exists between the incidence of scald and certain meteorological parameters during the period of cell enlargement of the fruit which occurs from 23 July until 3 September. The meteorological parameters which are closely correlated to scald incidence are total hours of sunshine and total rainfall during the period of cell enlargement. Information gained from this work allows a prediction to be made of the likely incidence of scald.

The prediction is made by determining the water deficit (evaporation minus rainfall) for the period 23 July to 3 September. Evaporation (E) can be determined by the formula:

$E = 47 \cdot 40 + 0.27s$ mm.

S = total sunshine in hours.

The total rainfall (R) in mm for the same period can be obtained from the nearest meteorological station and thus:

Water deficit = E–R.

The scald prediction can be assessed by the following table:

Water deficit (mm)	0 or less	0 to 50	50 to 127	127 to 203	203 to 255
Scald severity	nil	slight	moderate	severe	very severe

These predictions have been determined from experimental work with the variety, Edward VII, at the Ditton Laboratory and at East Malling Research Station. This method of forecasting is also likely to be relevant to other apple varieties that are scald susceptible. (Scald is also discussed in relation to spray materials on page 15.)

Effect of storage conditions

The severity of scald is much influenced by the conditions under which fruit is stored and on the time interval between harvesting and storage. The appropriate storage conditions should be established quickly after harvest. Fruit should be moved into store with minimum delay and the appropriate storage temperature and atmosphere for the variety achieved as quickly as possible.

The severity of scald is also influenced by the rate of loss of water from the fruit and fruit stored in bulk bins is more prone to the disorder because of the reduced water loss and the absence of adequate ventilation. Where ventilation of the fruit is increased by storage in bins with slatted bases and sides or by reducing the depth to which the fruit is packed into the container, scald incidence can be reduced.

Control

Scald can be well controlled by the pre-storage treatment of the fruit with the anti-oxidant ethoxyquin (Stopscald). This may be carried out by spraying the fruit before harvest or by dipping or drenching the fruit after harvest. The former has many limitations, not the least of which is that the spray must be applied on the day before harvest and good coverage is essential. This method of treatment is also wasteful of chemical. Dipping or drenching treatments after harvest are more effective and less wasteful.

Equipment for immersing bins or pallet loads of harvested fruit into a tank of ethoxyquin solution is relatively simple. It consists of a forklift truck fitted with an additional underslung fork assembly which is positioned slightly forward of the truck forks to allow clearance between the load and the sides of the tank. To prevent the fruit from floating the load is covered with a tarpaulin or plate which should be perforated to allow air to escape. The operation can be speeded by a hydraulic ram which moves the plate in and out of position. However, bins should be immersed slowly to avoid bruising the fruit which is pressed against the covering plate.

Alternatively, machinery for drenching fruit with ethoxyquin is available. The bins are loaded onto a conveyor belt and moved slowly under a cascade of ethoxyquin solution. Drenching units normally require a tractor mounted forklift as well as a forklift truck to operate satisfactorily, one to feed the bins onto the conveyor before drenching and one to remove the bins from the conveyor. Drenching units allow a greater throughput of fruit, and the control of scald is as effective as that obtained with total immersion. Ethoxyquin should be used at the following rates:

> For pre-harvest applications use 3·7 litres of ethoxyquin (70 per cent active ingredient) in 1000 litres of water.
> For post-harvest treatment use 2·4–3·7 litres of ethoxyquin (70 per cent a.i.) in 1000 litres of water.*

For post-harvest treatments the length of time the fruit should be immersed is dependent on the temperature of the solution and of the fruit. The rate of uptake of ethoxyquin by the fruit is higher at high temperatures. An immersion rate of 15 seconds is recommended for fruit temperatures of 16°C but this time should be increased to 40 seconds or more when temperatures are 10°C or below. Post-harvest treatments should be applied within seven days of harvesting, otherwise they may not be effective in controlling scald.

Ethoxyquin should not be applied to fruit of Golden Delicious because it produces a discoloration of the skin of this variety.

BITTER PIT

Symptoms

This is a disorder in which small areas of cortical tissue become brown and dry (see Plate II). These pockets of brown tissue are almost spherical and can be up to 5 mm in diameter. They usually occur just below the skin; the skin over the

---

* 2·4–3·7 litres/1000 litres ≡ 2·4–3·7 gal/1000 gal.

pits appears as a water soaked area and later becomes more highly coloured than the surrounding fruit surfaces. The cells of the damaged tissue gradually become dry and the skin over the affected area sinks leaving small round or angular, dark coloured pits in the skin. In severe cases the pockets of brown tissue can extend throughout the cortex of the fruit. The fruit may be affected whilst still on the tree or the disorder may only become apparent after a period of storage.

Control

Fertiliser practice should be adjusted so that excessive or irregular vegetative growth is avoided. Heavy dressings of potassium or nitrogen fertilisers should be avoided in orchards where bitter pit is known to occur (see page 12). It has been demonstrated that calcium deficiency in the fruit is perhaps the major factor associated with high levels of bitter pit. If measures to regulate the growth of the trees have failed to give good control of the disorder, they should be supplemented by foliar sprays containing calcium nitrate or calcium chloride. Sprays of calcium nitrate, 22 kg/ha in 2000 litres of water, or calcium chloride, 17 kg/ha in 2000 litres* with added wetter have given useful control of the disorder. Applications at low volume have generally been inferior to higher rates of application, but where only low volume spraying equipment is available, up to 16 kg/ha calcium nitrate or up to 13 kg/ha calcium chloride should be applied in the appropriate volume of water without wetter.

Spray timing

Timing of sprays differs with variety. For Worcester Pearmain or Merton Worcester, the first spray should be applied at the beginning of June but for Cox's Orange Pippin and other later maturing varieties the first spray should be delayed until mid to late June. Sprays should be applied at not more than 21 day intervals and should continue until harvest; the sprays applied between late July and the middle of September are particularly important.

Spray injury

Calcium chloride is as effective as the nitrate in controlling bitter pit but occasionally causes more damage to the leaves of sprayed trees. On the other hand, calcium nitrate may retard the development of red in coloured fruits and it may be best to use the chloride where maximum colour is required.

Calcium sprays are known to cause damage to the fruits of some varieties and Crispin, Bramley's Seedling and Discovery appear to be particularly prone. Others on which fruit damage has occurred include Ellison's Orange, Idared, Laxton's Superb, Lord Lambourne, Spartan, Ribston Pippin, Tydeman's Early Worcester, Merton Worcester and Newton Wonder. In many cases the damage was extremely light but it would be advisable not to spray Crispin and Discovery. Where Bramley's Seedling must be sprayed, half the recommended concentration of either the nitrate or the chloride should be applied every 14 days from mid-June to mid-September. For those varieties listed above and for some others

---

* 17–22 Kg/ha in 2000 litres ≡ 15–20 lb/acre in 175 gal.

the possibility of damage from the use of calcium sprays should be balanced against the anticipated severity of bitter pit. No fruit damage from calcium sprays has been recorded on Cox's Orange Pippin, Worcester Pearmain and Egremont Russet, all of which are susceptible to bitter pit.

Effect of storage conditions

Storage conditions also contribute to the prevalence of bitter pit. Any storage treatment which delays senescence will delay the development of the disorder and thus high temperatures, delayed storage and cooling, storage in air instead of in controlled atmospheres, will all accelerate the development of bitter pit.

LENTICEL BLOTCH PIT

This disorder is closely related to bitter pit but the damage occurs only to the skin of the fruit. Brown depressions or pits of up to 5 mm diameter occur around the lenticels. They are confined to the skin or peel and this separates the condition from bitter pit where the pits occur also in the cortex. Not infrequently both bitter pit and lenticel blotch pit occur together in the same fruit. The disorder is known to be associated with low levels of calcium in the peel of the fruit and this indicates a close relationship between lenticel blotch pit and bitter pit. Similar spray programmes with calcium as those for bitter pit control will reduce losses from this disorder.

LOW TEMPERATURE BREAKDOWN (LTB)

Symptoms

Breakdown of apples caused by low temperatures occurs in the cortical tissues as a general browning which also has a translucent appearance. The vascular tissues in the affected areas appear as dark brown spots. The damaged tissues extend to within a few millimetres of the skin of affected fruits but the skin, and the tissues immediately underlying the skin, generally remain healthy. Apparently healthy fruit may therefore be seriously affected internally. Symptoms of LTB in some varieties differ from those described above and take the form of a broad brown band around the circumference of the apple; the damage is much more superficial. This type of damage is sometimes described as ribbon scald.

Causes

While most commercially grown apple varieties are susceptible to LTB, low temperatures over a short period will not cause injury. Damage is more severe the lower the temperature and the longer the time the fruit is exposed to it. Losses due to LTB are never severe, so long as recommended temperatures are maintained, but injury can be expected in fruit from stores where the temperature has accidentally been reduced below recommended levels.

31

Symptoms

This disorder is associated with over-maturity of fruit. It is a progressive disorder which develops further when fruit is removed from store into higher temperatures (see Plate III). The early symptoms are variable and appear on the outside of the fruit as a general dullness of the skin, frequently seen first at the calyx. The skin then darkens in colour and breakdown of tissues gradually spreads into the cortex of the fruit. It therefore differs from typical low temperature breakdown which occurs internally before it spreads to the surface.

Causes

Senescent breakdown is more prevalent in late picked fruit and is more severe the higher the storage temperature. Storage in controlled atmospheres decreases its prevalence. There appears to be a relationship between low calcium concentration in the fruit and the prevalence of senescent breakdown and consequently calcium spray programmes similar to those described for the control of bitter pit may help to reduce losses. However, the main cause of the disorder is over-maturity due to late picking or prolonged storage, and these factors should be avoided.

CARBON DIOXIDE INJURY (BROWN HEART)

Symptoms

Brown heart is first recognised by a breakdown of the vascular tissues, but the symptoms are progressive and can increase to involve large areas of the cortex (see Plate VI). The injured areas are brown but generally firm and have a somewhat rubbery texture. In severe cases the damaged tissues lose water and large cork-like cavities appear.

Causes

Brown heart is caused by excessive levels of carbon dioxide (CO_2) accumulating in the store. Some varieties will tolerate much higher levels than others. Newton Wonder is very susceptible to injury and damage occurs in concentrations of CO_2 as low as 3 per cent. Conversely Worcester Pearmain can tolerate levels up to 13 per cent. Other factors may influence the prevalence of the disorder. For instance, susceptibility is higher when fruit is advanced in maturity and with some varieties sensitivity increases with low temperatures.

Another form of injury which does not resemble the symptoms described above can occur at low concentrations of carbon dioxide with low oxygen concentrations. The symptoms are very similar to those described for low temperature breakdown but the affected tissues are much firmer and have a hard texture. This type of injury has been observed in Cox's Orange Pippin when the oxygen concentration in a store, set for 5 per cent carbon dioxide plus 3 per cent oxygen at 3·5°C, had inadvertently been reduced to 2 per cent. This form of damage was especially prominent in fruit from those parts of the store where the temperatures were lowest.

Another form of carbon dioxide injury occurs occasionally on Bramley's Seedling and Edward VII. The damaged areas on the skin of the fruit are deeply sunken and are at first deep green in colour and progressively turn brown and then black. The margins of the damaged areas are irregular but sharply defined. This type of damage only occurs in controlled atmosphere stores and early in the storage period. Damaged fruit are generally found near to the walls and ceiling of the store and it is possible that, because of poor cold air distribution, they are cooled quickly whilst carbon dioxide is accumulating. The damage can be avoided if the fruit is cooled before the store is sealed.

INJURY FROM OXYGEN DEFICIENCY

Symptoms

If the oxygen concentration of a store falls below the recommended limits then damage to fruit may be caused by anaerobic conditions which produce alcohol poisoning. The symptoms can vary depending upon the variety and the duration of exposure to anaerobic conditions. The symptoms may be slight and show as surface indentations suggestive of bruising. Where symptoms are severe the skin of the fruit has brown patches with diffuse margins and damage can extend deeply into the cortical tissues which become pale brown. Affected fruit generally have an alcoholic or fermented odour and their taste is impaired.

Causes

Oxygen deficiency can generally be attributed to incorrect analysis by a faulty oxygen analyser or because the sample of gas analysed is not representative of the store as a whole. Store operators should use a portable analyser to check the gas mixture, especially in installations where all of the controlled atmosphere stores are sampled through long sampling lines, which may have developed faults.

WATER CORE

Symptoms

This disorder occurs in fruit whilst they are still on the tree and is not strictly a storage disorder. It can become less severe during storage but it is sometimes found in fruit removed from stores.

The disorder manifests itself as translucent or glassy areas of the flesh, but these areas are not always confined to the core of the fruit, but can occur anywhere in the flesh. Most frequently, however, they are close to the walls of the carpels or sometimes near vascular bundles.

Causes

The condition is generally associated with unbalanced tree growth, especially in trees with excessive vegetative growth. Orchard factors which are conducive to bitter pit are also conducive to water core.

There is some evidence that the condition is associated with low calcium levels in the fruit and if this evidence is confirmed it is possible that calcium sprays applied for bitter pit control will also control water core. The disorder occurs infrequently in the UK but has been recorded on the apple varieties Red Miller and Miller's Seedling. Fruit which is over-mature is the most likely to be affected and if the condition is seen, the fruit should be harvested quickly.

When the disorder is severe it can sometimes be followed by a complete breakdown of the flesh of the apple and this condition is sometimes referred to as water core breakdown, the symptoms being very similar to those described for senescent breakdown. Occasionally water core symptoms are still present in such fruit and the association between the disorders can be seen.

CORE FLUSH

Symptoms

Core flush sometimes appears late in the storage period of some apples especially Cox's Orange Pippin, Bramley's Seedling, Lord Lambourne, Laxton's Superb and Granny Smith. It manifests itself as a yellow-brown discoloration of the core between the carpels or seed cavities. The browning may involve all or part of the core area and sometimes extends beyond the core area into the surrounding flesh.

Cause

Core flush develops in fruit stored for long periods, and is more serious at low temperatures, is aggravated by high carbon dioxide concentrations, and increases if the water loss from the fruit is excessive. The incidence of the disorder is higher in fruit which is picked early.

FREEZING INJURY

Symptoms

If apple fruits are exposed to a temperature below 0°C their tissues become frozen. When the fruit is returned to a higher temperature the tissues thaw and collapse. This collapse of the internal tissues gives once-frozen fruit an irregular shape. Freezing injury is further characterised by cone-shaped areas of damaged tissue which have their apex at the core, and by the vascular strands within the fruit being light to dark brown. Severely frozen fruits become shrunken and wrinkled. When the tissues of frozen fruits are cut, juice oozes from the cut surfaces under slight pressure.

Fruit should not freeze in properly managed stores. Nevertheless this does occasionally happen, in which case frozen fruit may still be marketable but only if it is thawed out slowly. Rapid thawing invariably leads to heavy losses.

Storage diseases of apples

Losses of apples in store due to rotting by fungi can sometimes be very heavy. The extent of such loss is dependent on several factors, the most important of which is the presence of spores on the fruit surfaces, the quality of the fruit in terms of freedom from mechanical or other damage, the susceptibility of the variety to rotting and on the length and conditions of storage.

Some pathogens infect the fruit through the lenticels whilst others are wound parasites and can only infect through wounds in the skin of the fruit. Susceptibility to rotting gradually increases as the fruit ripens and invariably rotting is highest in fruit that has been over stored.

GLOEOSPORIUM

Symptoms and causes

This disease is caused by the fungi *Gloeosporium album* and *G. perennans*. It is the most important disease of dessert apples and if adequate measures are not applied, causes heavier losses than any of the other diseases. The fungi parasitise the wood of apple trees causing small cankers. Spores released from the cankers contaminate the fruit and infect through the lenticels or through cracks in the skin (see Plate V).

Symptoms of infection do not appear usually until after the fruit is stored. The rots produced by infection with *G. album* are fairly firm, circular and light brown in colour. Those produced by *G. perennans* are also circular, yellowish-brown in the centre, but with a dark brown perimeter. Spores of both fungi are produced on the surfaces of the rotted tissue during storage and can frequently be seen as a white translucent film over the rotted area. Losses from this disease are variable and depend on weather conditions before harvest. The disease is always most severe in wet seasons, particularly if high rainfall occurs in July, August and September. Further information on the relationship between the incidence of *Gloeosporium* and the nutrition of fruit trees is given on page 8.

Control

Losses from the diseases are minimised by protectant fungicide sprays applied before harvest, by post-harvest pre-storage fungicide treatments and by certain cultural practices. Cutting out the cankers from infected trees will reduce the spore inoculum in the orchard and so help to minimise the levels of fruit infection. Control by chemicals may be achieved by late summer orchard sprays of captan, benomyl, carbendazim, thiophanate-methyl or thiabendazole. These must be applied according to the manufacturers' recommendations.

Post-harvest immersion or drenching of fruit with benomyl, carbendazim, thiophanate-methyl or thiabendazole has been shown to give control of the disease. Such treatments can be applied by the equipment described for the application of ethoxyquin under apple scald (page 29). The chemicals should be used at concentrations and times recommended by the manufacturers. Post-harvest treatments should not be used as replacements for orchard sprays to control the disease but should be used to supplement the spray programme.

They are particularly useful on fruit from orchards where late spraying is difficult or impossible.

Symptoms and causes

Eye rot is caused by infection with the fungus *Nectria galligena*. The disease is prevalent in many orchards where it causes variable sized cankers on shoots, branches or even trunks of trees. Infection of the fruit may occur during fruit swelling and most commonly occurs at the calyx end or eye of the fruit. Eye rot is generally circular, flattened and with a dry dark brown appearance. However, other symptoms appear in some parts of the British Isles. Infection of Bramley's Seedling and some other varieties occurs through the lenticels and serious losses may result. This rot, caused by the same fungus, is generally slow growing, mid-brown in colour, and soft with a watery appearance.

Control

Infection of the fruit usually arises from spores derived from branch and shoot cankers and control of the disease on the fruit is consequential upon the control of the cankers in the orchard. This is best done by careful cutting out branch and shoot cankers or by painting them with fungicidal paints containing mercuric oxide, organomercury or 2-phenyl phenol.

Organomercury or wettable Bordeaux sprays applied at the beginning of leaf fall and again when approximately half the leaves have fallen have also reduced canker incidence in the orchard. Sprays of benomyl, carbendazim or thiophanate-methyl applied at frequent intervals through the spring and summer —mainly for the control of other apple diseases—have been shown to give a very considerable reduction in *Nectria* cankers.

Post-harvest immersion of fruit in benomyl or thiophanate-methyl will also reduce levels of fruit rotting caused by this disease.

BROWN ROT

Symptoms and causes

Brown rot is common on apples. It is caused by the fungus *Monilia fructigena* which is a wound parasite and can only infect the fruit through damaged areas of the skin, such as russet cracks, insect punctures or bird damage (see Plate IV).

The rot produced by the fungus is at first firm and mid-brown in colour and frequently the white mycelium of the fungus is apparent on the rotted surface. The rot spreads to involve the whole fruit and eventually the fruit becomes hard, mummified and almost black in colour. Development of the disease continues in store and spread by contact with adjacent, healthy fruit readily occurs.

Control

The disease is not easily controlled by orchard spray applications, although there is some evidence that sprays of benomyl applied at regular intervals through the growing season will give some control. Post-harvest immersion of fruit in benomyl or thiophanate-methyl will also reduce storage losses caused by this disease.

PENICILLIUM ROT

Symptoms and causes

Rots may be caused by several different species of *Penicillium* but the most common is *Penicillium expansum*. The fungus attacks through wounds or through lenticels and produces a characteristic greenish-brown, soft rot. The fungus is a common saprophyte but infection in store is mainly by contact between infected and healthy fruit. Losses are rarely serious enough to warrant the application of specific control measures but it is possible that post-harvest immersion or drenching in benomyl, carbendazim, thiophanate-methyl or thiabendazole will give some control of the disease.

PHYTOPHTHORA ROT

Symptoms and causes

Phytophthora syringae has become important in stored apples over the last few years and has caused very severe losses in some fruit stores.

The early symptoms of infection are a firm, brown rot occurring on the outside of the fruit. Usually circular in shape, the rot is dark brown in pears, and mid-brown in apples and extends into the flesh from the site of infection (see Plate VII). The skin over affected areas frequently appears marbled with brown and green. As fruit become wholly infected, a general khaki-coloured rotting occurs and this is accompanied by a vinegar-like smell and the rapid colonisation of the rotting tissue with secondary fungi, in particular *Botrytis*, *Penicillium* and *Rhizopus spp*. At this stage (reached by stored fruit in January–February) the *Phytophthora* rot is difficult to diagnose and can be confused with other fruit rots.

Infection occurs when soil or soil water containing spores comes into contact with the fruit; this may be by direct contact with the ground, by splashing during rainstorms or by mud contamination at harvest. If infected fruit is placed in store the disease can spread to adjacent fruit particularly during long-term storage.

Control

Post-harvest fungicide treatments are not effective against the disease. Control is best achieved by preventing infected fruits being placed in store and so care must be taken during picking to avoid mud contamination of fruit via baskets, bins and so on. Dropped fruit or fruit touching the ground should not be stored and if possible low hanging fruit should be picked separately from the remainder of the crop and not stored.

There are several other fungal diseases of stored apples, most of which are of minor importance. These are caused by *Botrytis cinerea* (see under Diseases of pears, page 39), *Phomopsis mali*, which produces brown lesions on the fruit late in the storage season, *Cladosporium herbarum* which frequently colonises apple fruits damaged by scald forming a black depressed rotted area confined to the scalded tissues, and *Trichothecium roseum* which is frequently associated with lesions caused by the apple scab fungus *Venturia inaequalis*. On apples these diseases are of little economic importance and do not normally justify special control measures.

Storage disorders of pears

Pears are prone to some of the physiological disorders which affect apples and their causes and the measures applied to control them are similar. The most important physiological disorders of pears are scald, superficial scald, freezing injury, core breakdown and skin damage caused by contact with packaging materials.

SCALD

Some varieties, particularly Williams' Bon Chrétien, sometimes show a brown discoloration of the skin which is initially superficial but which gradually progresses into the flesh of the fruit, especially if the temperature rises above that for normal storage. The disorder is most severe on immature fruits but is not controlled by pre-storage chemical treatments. Since the development of the disorder is favoured by temperatures higher than those recommended, it is essential to pick fruit at optimum maturity and store promptly at recommended temperatures.

SUPERFICIAL SCALD

Superficial scald is a disorder of Packham's Triumph and the American variety Anjou; it is characterised by brown to black discoloration of the skin. The disorder may show whilst the pears are in store but more frequently it appears only after the fruit has been removed from cold storage. The orchard conditions which predispose pears to this disorder are not well known but treatment of Anjou pears with ethoxyquin before storage is known to give good control and would possibly be an effective preventive. The effect of ethoxyquin treatment on stored Packham's Triumph pears is unknown but it is possible that such treatment would prevent the occurrence of scald in this variety also.

CORE BREAKDOWN

In core breakdown the core of the fruit goes soft and brown in colour and in severe cases the flesh is also affected. The disorder affects fruit which has been removed from normal storage temperatures to higher temperatures for ripening. The longer the fruit has been cold stored the shorter will be the time taken for core breakdown to appear after the fruit is removed from store. If pears are cold stored for too long they will develop core breakdown before they ripen. It is important not to over-store pears and to ensure the ripening process after storage is not carried out at too high a temperature.

FREEZING INJURY

The sympton of freezing injury on pears is similar to that described for the disorder under apples (page 34), but with pears cavities frequently appear in the damaged tissues of the flesh. The extent of losses from freezing injury is dependent on the temperatures at which the fruit is subsequently thawed. The lower the thawing temperature of the frozen fruit the fewer the losses.

INJURY FROM PACKING MATERIALS

Injury to the skin of pears is frequently caused by contact with the wood of the box or tray used for storage. The timber itself may cause injury or it may be caused by preservatives used on the timber. The injury takes the form of dark brown or black discoloration of the skin at the area of contact which is superficial and is usually confined to the skin. Timbers which are known to have caused damage include western arbor-vitae, *Thuja plicata* and Californian redwood, *Sequoia sempervirens*. Boxes or trays using such timbers in their construction should be avoided. Similar injury has been associated with fruit which has been in contact with wood treated with chlorinated naphthalene preservative.

Storage diseases of pears

The diseases affecting stored pears include *Gloeosporium perennans* and *G. album*, *Monilia fructigena*, *Penicillium expansum* and *Botrytis cinerea*. *Gloeosporium* species are much less important on pears that on apples, but a few isolated instances of severe losses from *G. perennans* have been reported on Conference in Kent. The most important diseases are those caused by *Botrytis cinerea* and *Monilia fructigena;* a description of the latter will be found under Diseases of apples (page 36).

BOTRYTIS ROT

Botrytis rot causes the greatest losses in stored pears. The fungus is a wound parasite and frequently invades the fruit tissues from the damaged stalk (see Plate VIII). The rot progresses rapidly into the fruit and is irregular in shape, dark brown or black in colour. The mycelium of the fungus is abundantly produced on the surface of the rotted tissues and this becomes a grey colour as the spores of the fungus are produced. This is especially noticeable under moist conditions. The rot will pass to healthy fruit in contact with affected fruit during storage.

The fungus is found on a wide range of moribund or dead plant tissue within pear orchards and infection of the fruit occurs through damaged tissues. To reduce the incidence of this disease it is important that fruit should be carefully picked to avoid excessive damage to the stalk and to avoid bruising. Any fruit with damage including that from pests and birds should not be stored. Immersion or drenching pears before storage in benomyl, carbendazim or thiophanate-methyl has given very good control of *Botrytis* rot.

Storage diseases of soft fruit

Rots caused by fungi are the most important cause of deterioration of soft fruit during storage. Cooling the fruit as soon as possible after harvesting to approximately 2°C and maintaining this temperature in an environment with a relative humidity of between 95–97 per cent will reduce storage rots to a minimum. Such an environment will reduce the rate of respiration of fruit and hence the loss of sugars and water. Water loss should be no more than five per cent during the storage period. Only sound fruit of top quality should be cool stored. Mechanical injury of fruit facilitates the development of spoilage fungi, and so it is vitally important that damage be kept to a minimum during picking and handling.

Experimental results from the Food Research Institute indicate that fruit picked in the latter part of the harvest period succumb to fungal decay more rapidly than fruit picked in the early part of the season. The relative importance of the different fungi in causing spoilage of stored soft fruit varies within a season as well as between seasons. It is also affected by the use of pre-harvest fungicide sprays.

BOTRYTIS

Symptoms and causes

Grey mould (*Botrytis cinerea*) is one of the most important causes of post-harvest rotting of all soft fruits. The symptoms are first a softening of the tissues under humid conditions. This is followed by a breakdown or rotting of the fruit which later becomes covered by the grey, spore producing bodies of the fungus. *Botrytis* rot of strawberries also causes brownish discoloration of the tissues. *The disease often attacks the flowers during wet or humid weather and remains latent within the developing fruit.* As the fruit matures the fungus grows and rots the ripening tissues. Although rotten fruit is not usually picked the fungus may develop after harvest in many fruits which appeared sound at the time of picking. The disease also spreads during storage from infected to healthy fruit by contact. The fungus is capable of growth at temperatures as low as 0°C, but storage at between 0–2°C will help to retard development.

Control

The incidence of the disease can be reduced by spraying with dichlofluanid, chlorothalonil, benomyl, thiophanate-methyl, carbendazim, thiram or captan. Sprays should be applied during the flowering period at the rate and times recommended by the manufacturers. Strains of *Botrytis cinerea* which are tolerant to benomyl, thiophanate-methyl and carbendazim have been found in strawberry, raspberry, blackberry and loganberry plantations and in these instances the application of such chemicals has been less effective in controlling the disease. In such circumstances growers should use alternative fungicides such as dichlofluanid. Dichlofluanid, however, is not recommended for use on protected strawberries because of the danger of leaf scorch. Fruit intended for canning, quick-freezing or jam, should not be sprayed with thiram or captan.

MUCOR MUCEDO

Symptoms and cause

Infection of soft fruits by *Mucor mucedo* results in severe softening of the tissues which eventually rot. The rotted parts of berries have a water soaked appearance, juice is exuded, and the fungus sporulates on the surface. It is initially visible as a greyish mass of spore containing bodies (sporangia) which turn black as they mature. The disease spreads during storage by contact between healthy and infected berries and is able to grow at 0°C. It has been most frequently observed on fruit harvested late in the season, though not on over ripe fruit, and where the incidence of grey mould has been reduced by fungicide sprays. This fungus is the predominant spoilage organism during storage.

Control

None of the fungicides at present in use are effective against *Mucor* sp. Prompt cooling and maintaining the fruit at 0–2°C will reduce the rate of development of the disease.

RHIZOPUS

Symptoms and cause

Rhizopus stolonifer (=*R. nigricans*) and *R. sexualis* cause similar symptoms on soft fruit to that of *Mucor mucedo* and it is only when the fungi are producing spores on the surface of rotted fruits that these organisms can be distinguished. The infection of a single berry by *Rhizopus* spp. can result in a whole punnet of fruit being rapidly covered with a mat of sporulating fungal mycelium. These fungi are most frequently observed on fruit harvested late in the season and on fruit where the incidence of grey mould has been reduced, but are not so frequently encountered as *Mucor mucedo*. 5°C is the minimum temperature at which *R. stolonifer* and *sexualis* grow.

Control

None of the fungicides presently available are effective against *Rhizopus* spp. but cooling and maintaining the fruit at between 0–2°C will help to prevent spoilage by these fungi.

Symptoms and cause

Cladosporium herbarum and *C. cladosporioides* can be a problem on stored raspberries and to a much less extent on blackberries and loganberries. The fungus grows on the surface of berries in a temperature as low as 0°C and is visible as a dark green mass of spores. Early stages of growth are greyish in colour and difficult to distinguish from grey mould.

Control

There are no label recommendations for chemical control of this disease, but in trials *Cladosporium spp.* were sensitive to dichlofluanid, benomyl, thiophanate-methyl and carbendazim but, as in the case of *Botrytis cinerea*, strains of the fungus tolerant to benomyl, thiophanate-methyl and carbendazim have been found in soft fruit plantations. Cooling and maintaining the fruit at 0–2°C will help to reduce the rate of development of the disease.

The storage of vegetables

The major requirements for successful vegetable storage are efficient control of temperature and humidity, both within the minimum allowed tolerances detailed in the section for specific crops, pages 47–54. The storage recommendations made should not be regarded as absolute, but only as safe limits within which various vegetables can be stored without deteriorating to a point at which they are unfit for sale or human consumption. These recommendations are based on the best experimental and commercial practices currently known but may be subject to change as more experimental data become available. Close adherence to the temperature and humidity levels recommended greatly enhances the potential storage life of the crop. Natural processes such as respiration continue after harvest and if storage conditions are not optimal, deterioration is likely to be more rapid.

Vegetables differ greatly in their storage potential, ranging from those of a leafy nature which are naturally short lived, like lettuce, to those which are naturally long lived, such as carrots and onions. Except in the case of onions, storage in no way improves the quality of the produce but merely extends its potential life. In practice two types of storage are in common use:

1. Short term—where the produce is held over a weekend or during periods of glut. This form of storage applies particularly to salad crops.
2. Long term—when produce is stored for as long as possible. This enables crops such as carrots, onions, winter white cabbage and red beet to be available throughout the year.

The approximate length of storage life to be expected in the absence of premature deterioration is given in detail later. Losses through deterioration while the produce is in store must be kept as low as possible, and quality must relate closely to that of the freshly harvested crop, particularly where stored and fresh produce compete in the same market. Storage losses can arise from three main causes:

1. The continuation of normal metabolic processes such as respiration, resulting in gradual change in composition which will eventually lead to over-maturity and death. These processes accelerate if produce is damaged in any way.
2. Dehydration.
3. The development of fungal or bacterial rots.

Many factors may contribute directly or indirectly to the length of life and quality of the crops in store. These include the method of growing, harvesting and handling, and store management.

Growing for storage

SOILS AND MANURING

Little is known about the effect of soil type but experience suggests that root crops from mineral soils store better than those from soils which have a high organic matter content. Some soil-borne diseases, such as *Sclerotinia* and *Centrospora* cause deterioration in store, particularly of root crops and celery kept for long periods. Cropping records should be kept and only those fields which have no past history of disease should be chosen for crop production. This may be difficult in cases where the disease has a wide host range or is not normally seen during the growing season.

The manurial status of the soil may influence storage life and the use of excessive nitrogen is known to be detrimental. Plants grown too soft are likely to suffer excessive mechanical damage during harvesting and handling, later resulting in the development of secondary rots in store.

CHOICE OF VARIETY AND LENGTH OF GROWING SEASON

There is practical evidence to show that for a number of different reasons some varieties of vegetables store better than others. For example late maturing varieties of celery store better than early varieties. Growing varieties known to be suitable for storage is a logical first step.

At present information is not available with which to judge the correct stage of maturity for the storage of vegetable crops. Field observations suggest that it is wise to harvest the crop before there is any indication of over-maturity; for example in celery the outer stalks become pithy when the heads begin to age and such stalks do not store well. Cauliflowers store best when the curds are still firm and tightly closed.

USE OF CHEMICALS ON THE CROP

From the wide range of chemicals used in vegetable production it is already possible to quote examples of pre-harvest treatments which influence the post-harvest life of stored crops. Desiccants, herbicides, fungicidal sprays and dusts, growth regulators, pesticides and nematicides must all be used carefully and growers should ensure that the materials they choose do not adversely affect the crop after it has been put in store. For example, an incorrectly timed desiccant spray applied to onion foliage before harvest can increase the amount of neck rot disease in store. On the other hand the beneficial effects of chemical treatment are also well known; the use of a benomyl seed dressing on onions can be directly linked with the good control of neck rot disease in store. Similarly, maleic hydrazide is applied before harvest to inhibit sprouting of onions in store.

Providing adequate routine control measures are taken during the growing season, damage from pests or diseases is not usually a problem. However, evidence suggests that carrot fly damage can be severe over winter even though routine control measures have been taken. Grower experience has established the advantage of avoiding this damage by early lifting and storage. Low levels of onion eelworm can result in breakdown in store and cauliflowers exhibiting severe symptoms of downy mildew (*Peronospora parasitica*) are not suitable for storage because of curd infection.

Harvesting, handling and preparation for storage

EFFECT OF WEATHER CONDITIONS

The onset of deterioration of salad crops and cauliflowers grown in high summer temperatures can be accelerated unless the crop is cooled quickly. Conversely crops harvested in autumn and winter, particularly brassicas such as winter white cabbage and cauliflower, must be stored before frost causes damage. Harvesting conditions during the autumn can vary considerably and where possible root crops should not be harvested in very wet weather. When carrots for instance, are lifted under such conditions, considerable amounts of soil, trash and disease inoculum are taken into store with the produce, which, apart from taking up valuable space, adversely affects the storage life of the roots.

HARVESTING

The optimum time for harvest can only be decided on the spot by the grower. Experience suggests that it is better to harvest crops a little early rather than late when there may be danger from bad weather, frost or over-maturity. The ideal method for long storage life is to harvest vegetables by hand but in practice this is often impossible, particularly with root crops. Several types of mechanical harvester are in use but whatever the lifting technique, every effort must be made to reduce damage to a minimum. Care at harvesting cannot be over-emphasised as only top quality produce should be stored. The after effects of damage on quality, storage and shelf life can be devastating. Some crops are more suited to mechanical harvesting than others and suffer less, but in almost every case prolonged survival after harvest depends on a low level of damage prior to storage. Types of damage vary, but it is important to realise that bruising, which may be difficult to detect in the freshly harvested crop, can be just as significant as visible cuts and abrasions. Recent surveys have shown the influence of machine setting and tractor speed on damage, and obviously the right adjustments for a particular situation can only be achieved by a skilled operator in the field.

HANDLING, TRIMMING AND PACKING

Physical damage can lead to accelerated deterioration whatever the crop involved. Crops such as winter white cabbage and cauliflower must be handled

45

individually; they should be placed, never thrown. The type of trimming required varies according to the crop, but it is important to follow the guidance given in the section that follows otherwise problems can arise from poor preparation. Washing prior to storage is not generally advised as this often leads to additional mechanical damage, and an increase in storage rots.

Vegetables in commercial stores are usually kept in bulk bins or containers stacked on pallet bases (see Plant and Equipment page 84). On-floor storage is sometimes practised in large units but care must be taken in handling both in and out of store.

Weight loss by dehydration in refrigerated cool stores can be reduced by lining the containers with polythene but the risk of certain diseases may be increased under these conditions. Dipping produce in a fungicide at harvest prior to storage is a technique which has shown promise in carrots and celery. However, some processing companies may have restrictions on the use of chemicals for crops to be processed, in which case, their guidance must be sought.

There are limitations on mixing crops in store, either because of differences in temperature and humidity requirements or because of the sensitivity of one crop to ethylene produced during storage by another. Top fruit and tomatoes should not be stored with vegetables: it has been shown that leafy vegetables become yellow and have a shorter storage life when stored with fruit. Apples can also cause flavour changes in carrots. In addition if cabbage, lettuce or cauliflower are in poor condition after a period in store, they tend to produce volatile substances which may have a deleterious effect on other crops.

Management in store

LOADING (SEE PLANT AND EQUIPMENT PAGE 103)

Field heat must be removed from vegetable produce so that it reaches the recommended storage temperature as quickly as possible after loading. The time taken to achieve this will depend upon the difference between the temperature of the produce and the store; also the heat output of the vegetables stored as the temperature is reduced. It is essential to know the temperature of the produce at all times and for this purpose temperature probes should be placed strategically in the produce for regular monitoring. Cases have been known where slow cooling from an inefficient cooling system during the early period of storage led to a total loss of the consignment. In the design of a new storage unit consideration should be given to the inclusion of a pre-cooling chamber to remove field heat prior to placing the crop in the permanent store; such a unit would be particularly useful where the holding stores are large.

CONTROLLED ATMOSPHERE STORAGE

Research into controlled atmosphere storage of vegetables shows benefit in terms of extended life and quality above that of refrigerated cool storage. The development and promotion of these findings is a slow process because controlled atmosphere stores are expensive to build and equip, and may not be

46

an economic proposition at present. Experimental work on this type of storage is in progress in the UK at Luddington Experimental Horticulture Station near Stratford-upon-Avon, Warwickshire.

MARKETING

If produce has stored well, preparation for market should be minimal. The cost of the storage process can be greatly increased if labour input is high in sorting and final preparation. The advantages of storage may be completely lost unless the produce is handled and marketed with great care after removal from store. Ideally transportation should be in temperature controlled vehicles. Temperature controlled cabinets are highly desirable to extend the shelf life at the point of retail sale.

In the following section the storage requirements for vegetables of major commercial importance are discussed: the most satisfactory temperature and relative humidity (RH) are given for long-term storage of each crop.

Dry bulb onions

Optimum storage conditions: Temperature 0°C
Relative humidity 70–80 per cent.

Onions can be stored in ambient ventilated stores from September until the end of March but for marketing during April–July, refrigerated storage is necessary. Onions held at 0°C can be marketed in excellent condition in June. For this to be achieved particular attention must be paid to the selection of bulbs for storage and to careful handling throughout. Onions should be selected from land known to be free of eelworm as infested bulbs are unsuitable for long-term storage.

To reduce sprouting in store the crop must be sprayed with a sprout suppressant prior to harvest. Maleic hydrazide has been approved under the Agricultural Chemicals Approval Scheme for this purpose. Harvesting operations ought to be completed by the middle of September. Later harvesting leads to a greater proportion of split and shed skins, also to the staining of onions left in the windrow. Onions intended for long-term storage must be harvested with the minimum amount of damage. Thick necked onions, weed trash, soil and waste bulbs should be removed during the loading operation. The elimination of soil during loading is especially important because soil cones in the stack lead to areas of dampness in which, after prolonged storage, re-rooting and shoot development can occur. Fast and efficient drying is essential. Drying must be continued until the necks of the bulbs are tight and dry.

Two systems of storage have been investigated at Kirton Experimental Horticulture Station in Lincolnshire. The first requires the use of cool storage facilities from September until July, the second from late December until July. In the first system the onions are dried in an insulated store in September and then in the first two weeks of November the store temperature is reduced to 0°C. Once the onions have been cooled it is necessary to keep them in that condition for the duration of the storage period. Humidity control is essential and although 70–75 per cent is optimal, it is both costly and difficult to achieve. Dry bulb onions will store well in relative humidities of between 75–80 per cent but above 85 per cent the bulbs develop roots and shoots making them unfit for market.

The second system requires the onions to be dried initially and maintained dry in an ambient ventilated store and then transferred to a refrigerated store in late December or early January. This method has practical advantages; the refrigerated stores can be used for other crops during September to January and an assessment of crop quality can be made in December so that only top quality bulbs are moved into the cool store. This allows greater flexibility than can be achieved if the cool store is used for both drying and storage. The main disadvantage is the double handling which increases labour costs and can add to the amount of mechanical damage.

Before removal from store the onions must be warmed up slowly, a few degrees per day, to a temperature above that of the dewpoint of the outside air; failing this moisture condenses on the surface. The process takes about five days (see Appendix 3).

In either storage system losses should not be greater than ten per cent.

Carrots

Optimum storage conditions: Temperature 0–1°C
Relative humidity 95–98 per cent.

The maximum storage life of carrots is unlikely to exceed 6–7 months. The principal aim of carrot storage is to enable buffer stocks to be available for marketing throughout the season and to provide carrots capable of competing with high price, imported roots during the months of May and June. The outlets, depending on quality, include the fresh market and processing industry. Ideally, carrots should be grown for storage and not, as so often happens in practice, be put into store merely because they happen to be available and surplus to immediate requirements.

For long-term storage hand lifting is the only method recommended. Mechanically harvested produce of good quality may be acceptable but only if the roots are required before mid-March. Beyond this date, a rapid breakdown can occur in some seasons, due primarily to mechanical damage and subsequent disease infection. This breakdown may be alleviated by a benomyl fungicide dip prior to storage.

It is important that carrots be always harvested under the best possible soil and weather conditions. Repeated trials have shown that produce harvested in cold wet weather, with large amounts of soil and debris adhering to the roots, does not keep so well as crops lifted under dry conditions. Great care should be taken to minimise damage during harvesting. Severely damaged carrots must not to be used for long-term storage.

A recent survey during lifting and preparation, showed that up to 58 per cent of roots were damaged, of which 29 per cent was attributed to the harvester. The damage recorded when two operators of different skills worked on the same harvester was 18 per cent in one case and 60 per cent in the other, showing the value of the skilled harvester operator. Considerable variations can occur between varieties, for example, Autumn King and Chantenay fracture less easily than Amsterdam or Nantes.

Carrots are usually stored in bulk containers, of either 500 kg or 1000 kg capacity. Storage in sacks is not recommended because the airflow is restricted making it difficult to keep the roots cool in store. The main problems encountered within the store are dehydration, often caused by incorrect airflows, and disease. In present day commercial stores it may be difficult to maintain the relative humidity at 95 per cent, but it is known that weight losses above eight per cent through dehydration, materially affect crop quality.

ADAS trials in East Anglia and at Kirton Experimental Horticulture Station in Lincolnshire, have shown that polythene lined bins, used in combination with roots dipped in a fungicide, gave the best control of dehydration and major storage diseases. The diseases involved and the dipping technique employed are described on page 56.

Red beet

Optimum storage conditions: Temperature 3°C
Relative humidity 95–98 per cent.

Red beet can be stored for six to eight months in temperature controlled stores. It can be kept in clamps or barn stored during the earlier part of the winter and moved later into temperature controlled chambers as outdoor temperatures rise.

Red beet is harvested during November before severe frosts occur, great care being taken to avoid mechanical damage. The beet are topped and sorted to remove poor quality, diseased or damaged roots before loading into bulk bins. Regrading will be necessary on removal from store.

Lining bulk bins with 500 gauge polythene improves the quality of roots compared with those stored in unlined bins and reduces weight loss through dehydration. Care must be taken not to expose red beet to temperatures below 0°C as under these circumstances sugar conversion can be initiated with a subsequent loss of flavour.

Celery

Optimum storage conditions: Temperature 0·5–1°C
Relative humidity 95 per cent.

The maximum storage life of celery varies from 12–14 weeks in the absence of disease. Where disease is present breakdown usually occurs after about six weeks in store. It is wise to consider for storage, only celery from fields which have not previously been intensively cropped with celery or carrots. Crops from fields with a past history of soil borne disease must not be stored. Celery should always be grown in a wide rotation with unrelated crops.

Celery ought to be harvested before severe or continuous frosts occur. Late maturing varieties are preferable to self-blanching types as the latter tend to be pithy late in the season. There is some evidence that there may be varietal differences in sensitivity to low temperatures within the store. For example FMD5 appears to show a greater liability to damage.

Crops intended for storage are usually mechanically undercut, hand lifted and trimmed in the field. The sticks of celery are then packed individually into plastic containers (wood and fibreboard containers may lead to excessive damage followed by subsequent rotting). To reduce the risk of damage during stacking in store, excess foliage is trimmed off to the level of the plastic container. It is not recommended that celery be stored as a jumble pack in bulk bins.

A benomyl dip of the base of the trimmed sticks is recommended for the control of *Centrospora* rot and *Botrytis*. Details of this treatment are given on page 59. Lining the bin with 500 gauge polythene reduces losses from dehydration.

Only top quality celery can be sold on the fresh market. Celery which does not meet these requirements is used for processing as soup or for canning.

Winter white cabbage

Optimum storage conditions: Temperature 0°C
Relative humidity 95 per cent.

Cabbages can be stored for up to eight months in cool stores with a weight loss averaging $1-1\frac{1}{2}$ per cent per month and total wastage, including trimming, of 20–25 per cent. It is essential that such cabbage should be harvested in advance of severe or prolonged frost. This means that in the UK the crop must be in store by early to mid November.

Extreme care must be taken during all stages of harvesting to ensure that the heads reach the store free from any form of physical damage. The cabbages should be cut cleanly with a sharp knife close to the base, so that all loose outer leaves fall away leaving a dense head with a cleanly cut stem about 5–10 mm long. Any discoloured or damaged wrapper leaves need to be peeled off and the heads carefully placed into bulk boxes or trailers for transport to the store. The cabbages may be stacked up to 3 m high or alternatively, stored in bulk bins. Once in store the cabbages should be held at a temperature of 0°C with a relative humidity of 95 per cent.

Re-sorting may be necessary during the storage period and diseased heads will need retrimming. Disease free heads should not be peeled until required for market.

Peeling to remove senescent leaves and those infected with *Botrytis* or bacterial soft rots, and butt trimming will then be necessary. On average a total of four to six wrapper leaves are removed.

Cauliflower

Optimum storage conditions: Temperature 1°C
Relative humidity 95–98 per cent.

Cauliflowers can be stored at 1°C for periods of up to three weeks. Storage beyond this point leads to rapid deterioration and loss of shelf life.

Work at Kirton and Luddington Experimental Horticulture Stations has shown that only slightly immature Extra and Class I heads should be used for storage. Sufficient outer leaf should be left to cover the curd and allow for further trimming when the heads are removed from store. Extreme care must be taken during the harvesting operation to ensure that the heads reach the store as free as possible from any form of physical damage. Bruises and cuts to the curd turn brown during storage and are quickly colonised by such soft rotting organisms as *Erwinia sp.* and *Pseudomonas spp.* Heads showing leaf scorch are unsuitable for storage.

Once the heads are harvested the field heat should be removed as quickly as possible. It is an advantage to cut cauliflowers early in the morning when it is cool as this enables the required temperature of 1°C to be reached with the minimum of delay. The longer it takes to reduce the temperature the shorter the storage life of the crop. The ice bank cooler using positive ventilation is ideal for the rapid cooling of cauliflowers.

The choice of container will depend on several factors; the duration of storage, the type of market outlet and the cooling facilities that are available. Both wooden crates and 500 kg slatted wooden boxes have been used success-fully but cauliflowers are difficult to cool in bulk. If the temperature in the centre of a bulk box cannot be reduced below 4°C within the required period, smaller containers should be used. If the crop is to be stored short term, it is satisfactory to use wooden market containers, but for longer periods of storage during which there will be some deterioration and some butt discoloration it is better to store in bulk boxes. Cauliflowers intended for pre-packing are best stored in bulk boxes. Fibreboard boxes are unsuitable as storage containers.

The main difficulty encountered when storing cauliflowers, as with any leafy vegetable, is the rapid loss of moisture that occurs. When this exceeds five per cent the outer leaves wilt and the curd becomes rubbery. Nevertheless, weight loss can be minimised by maintaining humidities in the region of 95–98 per cent. Many stores on commercial holdings, especially where doors are frequently opened or closed for the loading or removal of produce, do not achieve the high humidity conditions essential for the successful storage of this crop. Polythene pallet covers placed over crates or boxes of cauliflowers which have been pre-cooled, help considerably to reduce weight loss. Also the high humidities obtained in an ice bank store are ideal for preventing moisture loss.
Short-term storage: 1–7 days.
As quickly as possible cool the cauliflowers to 3–4°C at which temperature they may be held, though 1°C is better.

Long-term storage: 7–21 days.

Cool the cauliflowers to 1°C as quickly as possible and in conventional stores the pallets should be enclosed in polythene covers to reduce desiccation. A gap of 100–150 mm must be left between the pallets to allow adequate air circulation for the maintenance of the required temperature within the containers. In the ice bank store this covering will not be required. After two weeks storage retrimming will be necessary and after three weeks retrimming and regrading will be required.

Brussels sprouts

Optimum storage conditions: Temperature 0°C
Relative humidity 95–98 per cent.

Brussels sprouts deteriorate very rapidly at temperatures above 10°C. Loss of quality is characterised by yellowing of the outer leaves, loss of turgidity and browning of the cut ends. Generally under commercial conditions storage life will not extend beyond seven days; however, maximum storage life depends to a large extent on the rapidity of cooling and if fast effective cooling can be achieved storage life can be extended. Good air circulation through the container is essential if rapid cooling is to be achieved. The ice bank cooling system with positive ventilation is ideally suited for this. In a conventional store the smaller the container the faster generally the cooling rate.

High humidities are essential for the successful storage of this crop but a loss of moisture through transpiration tends to be high even if the relative humidity is kept at the recommended level.

Lettuce

Optimum storage conditions: Temperature 0·5–1°C
Relative humidity 95 per cent.

Lettuces are highly perishable and should therefore be stored only in periods of surplus and over weekends or holidays. Deterioration shows as wilting, yellowing of the outer leaves and discoloration of butt ends. Generally, storage beyond seven days cannot be recommended.

Successful storage largely depends on prompt cooling after harvest. For this reason vacuum cooling has been found to be the most effective method available for pre-cooling lettuce. So much so that in the USA, where greater distances and temperatures are involved, the majority of lettuce is packed into cartons and is vacuum cooled immediately before shipment in refrigerated transport.

Storage at 0·5–1°C greatly retards deterioration, but as lettuce is easily damaged by freezing, it is important that all parts of the storage room be kept above freezing point.

High humidity is essential to maintain lettuce in a fresh turgid condition. For this reason individual head wraps using perforated plastic bags help to reduce weight loss during storage.

Salad onions

Optimum storage conditions: Temperature 0°C
Relative humidity 95–98 per cent.

Salad onions should preferably be stored in units no larger than 5–7 kg. After storage at 0°C and at a relative humidity of 98 per cent a shelf life of three days can be expected. For successful storage it is essential for the field heat to be removed quickly, certainly within 3–4 hours. For this purpose hydro-cooling is recommended.

Asparagus

Optimum storage conditions: Temperature 0–2°C
Relative humidity 95 per cent.

Asparagus can be held in store for short periods of up to ten days. Even at 0°C the spears slowly lignify and lose flavour, although deterioration is only just noticeable during the first week. At higher temperatures deterioration is more rapid and above 3°C the spears will continue to grow and bend upwards if kept for more than a few hours.

Initial cooling and extraction of field heat will be facilitated if the boxes of spears are stacked so that air can circulate freely around and through each individual stack. After initial cooling, the boxes or stacks are covered with polythene to prevent dehydration during the remaining period of storage.

Asparagus spears can be injured by exposure to excessively low temperatures, precise temperature control is therefore essential.

Tomatoes

Optimum storage conditions: Temperature 7·5–8·0°C
Relative humidity 85 per cent.

At 7·5 to 8°C tomatoes may be held in store successfully for up to ten days. A relative humidity higher than 85 per cent may increase the likelihood of grey mould (*Botrytis*), especially on the calyx.

Tomatoes, being of tropical or sub-tropical origin, require storage temperatures considerably higher than most indigenous fruit and vegetables. At temperatures below 7°C symptoms of injury from chilling will develop, either in store or during subsequent ripening at ambient temperatures. The extent of the injury will depend on the temperature and duration of storage. Slight chilling is characterised by changes in pigmentation of the skin which develops a somewhat orange hue, in contrast to the normal, deep red colour. Severe injury will occur at 2°C where the symptoms are rugose or even pitted skins, blotchy, uneven coloration with a translucent appearance; ultimately a large proportion develop bladdery, water soaked areas. In general, the riper the fruit the less susceptible it becomes to injury.

The following procedure is recommended for the successful storage of tomatoes. The fruit must be selected at the stage of one-quarter ripe since some colouring up will occur during storage. After picking, over-ripe, damaged or diseased fruit must then be culled and fruit that is free of blemish transferred to shallow stacking boxes or baskets arranged on pallet bases. The containers can then be moved to the storage room and the temperature set at 7·5 to 8·0°C. The time interval between picking the fruit and reaching the store must be kept to the minimum. When required for market, the fruit should be transferred to a warm, dry, well-ventilated room or passage where condensation will disperse within two to three hours. When dry, the fruit is ready for grading and packing into retail or wholesale containers.

Potatoes (for marketing as new)

Optimum storage conditions: Temperature 4–5°C
Relative humidity 95 per cent.

Only very limited quantities of immature potatoes for canning as new are held in refrigerated stores. Storage is usually required for short periods of about three months but can be particularly valuable in allowing the canning of potatoes on processing lines used for other crops and for helping to level out irregular supplies.

Potatoes for canning must be kept at a low temperature to maintain their new condition and so, after a curing period, they should be held at 4–5°C. Since new potatoes are characteristically sweet, the sweetening associated with low temperature storage is not a disadvantage. Humidities in the region of 95 per cent are required.

Storage diseases and disorders of vegetables

Crop production, careful harvesting, and store management have been discussed in a previous chapter. These factors are of paramount importance in achieving successful storage and contribute to the control of diseases in store. Neglect of the basic principles of good husbandry and store management will lead to increased problems from attack by fungi, bacteria and viruses in the various crops. In some cases fungicides are available which will assist control of certain storage diseases. The following section deals with particular crops and outlines their major problems.

Dry bulb onions

NECK ROT

Symptoms and causes

This disease, caused by the fungus *Botrytis allii* can lead to severe losses in stored onions. The necks and tops of affected bulbs become soft and discoloured and eventually a grey mould may develop on the neck in which black bodies or sclerotia can be seen (see Plate IX). As the disease is seed borne, infection of the young seedlings can take place from contaminated seed samples, but spread may also take place later in the field at any time during the growing season. Normally there are no leaf symptoms. Experiments indicate that the earlier infection takes place, the more serious the disease in store, although external symptoms are rarely seen in the freshly harvested bulbs. Topping very near the neck before harvest or excessive mechanical damage during the harvesting process tends to accentuate the problem in store. Inefficient drying may also allow latent infections to develop in the neck.

Control

A benomyl seed dressing is recommended for the control of the seed borne phase of this disease.

As *Botrytis* will colonise broken tissue onions must be carefully handled at harvest, in the windrows and during removal to the store. The technique used for drying onions successfully is described in Horticultural Enterprise Booklet 1—*Dry Bulb Onions*. Although infection normally takes place in the field, neck rot outbreaks in store will be more severe if the crop is not dried well and if it is not kept dry until it is moved into refrigerated storage.

Bulb staining, which can spoil the appearance of cleaned onions prepared for market, develops as a result of a prolonged period in contact with wet soil, and is not due to storage conditions. Staining is more severe when bulbs are damaged mechanically during harvesting operations.

Carrots

GREY MOULD

Symptoms and cause

Infection by the fungus *Botrytis cinerea* can cause a great deal of trouble in bulk bins of carrots, particularly when they are damaged and lifted under bad weather conditions. Infected roots develop light brown lesions which have a water soaked appearance. Eventually they become covered with a grey mould growth, sometimes accompanied by small black bodies pressed into the decaying tissues. The roots do not collapse, but as the decay advances they become leathery. The fungus is a weak parasite which can attack damaged roots more easily then perfect ones. It can be found everywhere but particularly on moribund plant material; so if much leaf debris and trash are mixed with the crop in store the roots may soon become diseased. Once grey mould develops, spread can be by contact or by air currents carrying spores from infected roots to other parts of the store. Results of recent experiments show that the disease is more widespread in samples suffering from dehydration. Grey mould may appear earlier in the storage period than other storage rots, particularly if temperatures are allowed to rise, or the roots are not turgid.

Control

Care in harvesting and handling the crop before storage will reduce both mechanical damage and the amount of debris in the bulk. Roots should be cooled quickly and the relative humidity kept as high as possible to maintain them in a turgid condition. Lining bins with polythene is one way of achieving this.

A fungicide dip of benomyl at 500 g active ingredient (a.i.) in 1000 litres of water at lifting has given control of this disease. Bulk bins can be dipped in a large tank containing the diluted fungicide so that the solution reaches every root The bins are then removed and allowed to drain before loading into store.

LICORICE ROT

Symptoms and cause

This disease is caused by the fungus *Mycocentrospora acerina* which, like grey mould, has a wide host range including other vegetables such as celery and parsnips (see Plate X). However, unlike grey mould, it is most frequently soil borne. Root symptoms are not normally seen in the field before harvest, although the fungus has been reported occasionally on foliage. A dry black rot often develops at either the crown or the tip of a root although side lesions also occur. The lesion itself is black with a brown, advancing margin which eventually penetrates deep into the root. No fructifications are visible on the surface of these black lesions and normally affected roots remain fairly dry. Infection takes place during storage on the root where the fungus is in contact with the surface. Roots seem to be resistant in the early part of storage and it is not until they have been stored for three or four months that the majority of lesions develop. Infection does not spread from root to root by contact.

Control

The best control is achieved by selecting fields free from the disease, and these can be found, even in intensive carrot growing areas. Undamaged roots are less likely to be infected and therefore mechanical damage should be kept at a low level.

A fungicide dip of benomyl at 500 g a.i. in 1000 litres* of water at lifting has given control of this disease.

CRATER ROT

Symptoms and cause

The fungus *Rhizoctonia carotae* is the cause of this storage rot in carrots. Under dry conditions small pitted spots containing a white growth develop on the surface (see Plate XI). When this growth is examined under the microscope, fungal threads with a characteristic crystalline appearance can be seen. The spots enlarge to form craters below which the decayed tissue is brown, firm and dry. In conditions of very high humidity, a greyish white web of fungal threads spread rapidly so that the containers and large numbers of roots are covered.

Control

This disease is associated with contaminated containers and samples containing too much rubbish and soil. Spread takes place most rapidly when humidity is very high, and although bin lining may reduce grey mould, the high humidities which develop in the lined bins may assist the spread of crater rot. Bins, while not in use, should be sterilised with a suitable fungicide such as formaldehyde.

SCLEROTINIA ROT

Symptoms and cause

The name, cottony soft rot, sometimes used in America to describe the disease caused by the fungus *Sclerotinia sclerotiorum*, gives an excellent picture of the symptoms of this disease. Unlike licorice rot these symptoms may sometimes be detected in the field. Infected roots become covered with a dense white cottony growth in which large black bodies (sclerotia) may be embedded. The infected tissue remains a carrot colour but gradually softens and collapses. The

---

*500 g/1000 litres $\equiv$ 1lb/200 gal.

fungus spreads from infected roots to others in contact with them so that nests of roots can be seen within the bulk of a container, particularly if moisture condenses on them. The collapse of infected tissues gives rise to liquid which may contaminate others, and secondary bacterial decay may follow. The fungus has a wide host range among vegetable crops and the sclerotia are able to survive in soil from season to season. When lifting it is likely that soil carrying infection, or infected roots themselves may be lifted with healthy ones and from these sources spread of the disease will soon start in the store. This fungus is less able to grow at low temperatures than *Rhizoctonia carotae* or *Botrytis cinerea*.

Control

Fields with a history of sclerotinia rot should be avoided as no root sorting is possible in machine lifted samples. Spread within the bins is favoured by very high humidity but a dip in benomyl at 500 g a.i. in 1000 litres of water at lifting, gives some control of the disease, even in polythene lined bins.

BACTERIAL SOFT ROTS

Collapse of carrots stored in large containers can occur after the development of soft rots caused by bacterial infection from organisms such as *Erwinia carotovora* or *Pseudomonas* spp. Mechanical or any other type of damage should be avoided as much as possible to reduce the exposure of internal tissues to bacterial infection. These rots may also be a problem during the shelf life period if the roots are washed and packed in plastic bags.

STORAGE DISORDERS

Carrots can suffer severe dehydration which may lead to problems with grey mould, but in polythene-lined bins there is a tendency for shoot and root development to take place thus lowering the quality of the roots. Carrots should not be allowed to lose more than eight per cent by weight of moisture. To avoid development of a bitter flavour carrots should not be stored with or near fruit such as apples or pears, which produce ethylene.

Red beet

DRY ROT

Symptoms and cause

This is caused by the fungus *Phoma betae*. On any part of the root surface, sunken black lesions develop causing the flesh to turn reddish-brown or black (see Plate XII). The disease can be seed or soil borne and is sometimes associated with fields which are low in boron.

Control

This is best achieved by sowing seed that has been given the thiram soak treatment. Crops should be grown in fields which have not had close rotations of red or sugar beet.

The maximum permissible water loss for beet is ten per cent and dehydration may occur particularly when the crop is damaged. Where the roots have suffered from over heating or freezing the affected tissue is a darker colour than normal and oozes deep red droplets of sap when cut. Beet can suffer from carbon dioxide injury when the level is above five per cent in the store atmosphere. Regrowth of the tops may occur in very humid conditions.

Celery

GREY MOULD

Symptoms and cause

The fungus *Botrytis cinerea* can infect the leaves or stalks of celery in store, particularly if the plants are not turgid and there is mechanical damage or abrasion. Affected areas appear water soaked and the typical grey mould growth develops in favourable conditions.

Control

Adequate trimming to remove damaged outer stalks and as much leaf as possible will help to prevent infection. The produce should be cooled rapidly after lifting and the relative humidity maintained to reduce water loss which should not be more than six per cent.

CENTROSPORA ROT

Symptoms and cause

The fungus *Mycocentrospora acerina* mentioned under carrots also infects celery causing a greenish-black rot at the junction of the butt and stalks (see Plate XIII). The disease is not normally seen in the field or at harvest, but symptoms develop in store after the celery has been kept for six to eight weeks. In severe outbreaks the rot development is rapid after this time. Each stick is infected individually from the disease inoculum introduced into the store with the crop. Fields free from this soil borne disease should be chosen for the stored crop.

Control

Centrospora rot can be controlled by dipping the base of the trimmed sticks in benomyl. Packed crates of celery can be passed through a benomyl solution of 500 g a.i. in 1000 litres* of water at harvest, allowed to drain for a short period and stacked in bulk bins for storage. Where no treatment is given the crop should be examined for symptoms about seven weeks after storage and if these are found, the crop should be marketed quickly to avoid further loss.

---

* 500 g/1000 litres ≡ 1 lb/200 gal.

E

Symptoms and cause

Bacterial soft rot may attack celery both in the field and during marketing. The causative bacteria are probably *Erwinia carotovora* and species of *Pseudomonas*. Infection usually occurs through wounds or through lesions caused by other diseases. Decay in the leaf stalks may begin as small, water soaked areas; as the spots enlarge the infection penetrates deeper and spreads along the stalk. The initial water soaked appearance then changes to a pale brown colour and the decayed tissue becomes soft and wet. Washing water which has become heavily contaminated with bacteria may be a source of inoculum, and spoilage is favoured by maintaining celery in a moist condition, especially in plastic bags, without refrigeration.

Control

Storage at low temperature 0–1°C will minimise spoilage from bacteria.

STORAGE DISORDERS

Pithiness is a physiological disorder which may start in the outer stalks in the field and increase in store. The central tissue of the stalk becomes soft and spongy and later collapses. Early maturing varieties, such as the self blanching types, are less suitable for long-term storage because of pithiness. The condition may also occur in all varieties if plant growth is temporarily checked by such factors as water shortage.

Winter white cabbage

GREY MOULD

Symptoms and cause

Grey mould is commonly encountered in stored cabbage. Infection by the fungus *Botrytis cinerea* can cause loss directly from rotting of the outer leaves and indirectly from increased costs of time and labour needed to trim the heads before marketing. *Botrytis* is typically a wound parasite and therefore symptoms are not normally seen at harvest although the fungus, which has a very wide host range, is everywhere. Brown areas appear first on the outer wrapper leaves and gradually the fungus growth develops over much of the surface like a blanket. As the outer leaves senesce they tend to lose their normal green colour and become flaccid. Under these conditions *Botrytis* spores from infected heads spread to other cabbages giving rise to further infections. In the later stages of *Botrytis* breakdown, secondary bacterial soft rots with associated sliminess often follow, making the cabbages unpleasant to handle.

Control

The crop should be harvested before frosts occur and loose outer leaves trimmed. Because infection takes place rapidly where plant sap is present great care is necessary in handling each cabbage so that the wrapper leaves are not broken or bruised, High humidity in the store should be maintained so that the heads keep turgid. Frequent trimming is not necessarily an advantage in the prevention of this disease as at each trimming, clouds of spores will spread from the mouldy cabbages to freshly trimmed surfaces and quickly start new infections. While the cabbages are left undisturbed spore dispersal will be slight.

PHYTOPHTHORA ROT

Symptoms and cause

This disease is most likely to develop when cabbages are harvested in very wet field conditions. The organism *Phytophthora porri* can be carried from the soil by rain splash on to the head or cut stalk. The outer wrapper leaves of infected cabbages are yellow or brown in colour. When the heads are cut open a dark brown discoloration can be seen spreading upwards from the cut stem base into the leaves (see Plate XIV). In severely affected cabbage the entire head may become discoloured, but the tissue remains firm.

Control

The crop should be harvested under dry conditions directly into boxes, and not left on the ground or in the field exposed to rain splash. Stalk trimming during the mid storage period should be avoided. No control by fungicide is available.

BACTERIAL SOFT ROT

Symptoms and cause

Bacterial soft rots of cabbage and other vegetables are caused either by *Erwinia carotovora* or by species of *Pseudomonas*. The bacteria are commonly found in association with the crop and invade plant tissue through wounds, through lesions produced by other diseases or through tissue damaged by freezing. Symptoms usually show at first as water soaked areas on the leaves; these extend until a large proportion of the outer part of the head is converted to a brown, slimy mass.

Control

Storage at low temperature (0–1°C) will slow down the rate of spoilage in comparison with rot development at room temperature. Care should be taken

Cauliflower

BACTERIAL SOFT ROT

Symptoms and cause

The bacteria *Erwinia carotovora* and *Pseudomonas* spp. are commonly found in association with the crop and invade plant tissue through wounds to which cauliflower curds are most susceptible. Infection begins as small, yellow water-soaked specks in the florets, and as decay progresses the florets become increasingly discoloured and water soaked. Infection probably follows pressure bruising in handling and packing.

Control

Storage at low temperature (0–1°C) is essential and curd damage must be avoided.

STORAGE DISORDERS

The cauliflower curd is a delicate structure which will suffer from any form of rough handling. If the floret tissues are broken, surface micro-organisms will multiply in the released sap and as a result discoloration may occur. Quality loss is also linked with leaf yellowing and poor butt colour. After one week the butt looks dry and subsequently becomes progressively discoloured so that re-trimming is essential. A rubbery texture of the curd indicates dehydration.

Brussels sprouts

BACTERIAL SOFT ROT

Symptoms and cause

This is caused by *Erwinia carotovora* and by species of *Pseudomonas* which are commonly found in association with the crop and invade through wounds or tissue damaged by frost. The symptoms may show as water soaked areas, or the tissue may become discoloured and black.

Control

Cooling and maintaining a temperature of 0–1°C will help to reduce the occurrence of spoilage.

Variations in keeping quality from season to season have been noticed in commercial storage. The main problems are dehydration, leaf yellowing and butt discoloration. Water loss greater than six per cent leads to unacceptable dehydration; so high humidity must be maintained throughout the storage period. No measures are yet known that will delay the onset of leaf yellowing and butt discoloration.

Lettuce

Symptoms and cause

This may cause serious losses during marketing. The causative bacteria are thought to be mainly species of *Pseudomonas* and occasionally *Erwinia carotovora*. Decay may start on bruised leaves or at the leaf margins following tip burn or marginal browning, or may follow other diseases or freezing injury. Infected leaves at first appear water soaked and later become discoloured and slimy.

Control

Bacterial soft rot can be minimised by avoiding storage of lettuce showing tip burn, by trimming to eliminate damaged or diseased leaves and by maintaining the crop at a low temperature (0–1°C) during storage. The butt often discolours after 24 hours.

STORAGE DISORDERS

If the leaves lose turgidity, wilting and *Botrytis* infection follow, resulting in quality loss and a shorter shelf life. Water loss should not exceed five per cent.

Other vegetables

Successful storage of tomatoes, salad onions and asparagus depends on maintaining the optimal conditions described on page 53. Fungal disease are of little significance.

Layout, design and construction of stores

It is important to define the function and uses of a proposed refrigerated cool store before the planning starts so that the designer and refrigeration engineers can determine the performance expected from plant and the standard of insulation required in order to achieve the specified conditions. The main considerations are:

Types and varieties of crop to be stored.
Duration of storage.
Scale of enterprise and over-all turnover.
Methods of handling crops and of filling and emptying stores.
Possible alternative uses.

As an alternative to constructing a new store, conversions of existing buildings should be considered. However, such buildings will often be found unsuitable due to narrow widths, poor access and low eaves, or will be found after survey to be structurally unsound. More modern structures will sometimes prove suitable, particularly for housing free standing stores.

At an early stage, preliminary cost estimates and a feasibility study will indicate whether in economic terms it is worth proceeding with the project. Assuming this is so and that a suitable site is available, sketch plans can be drawn and the store design agreed. For guidance a number of store types are illustrated, with a range of constructional details and methods, including the use of insulation materials in relation to new layouts. It is not possible to discuss conversions in this book but the same principles of store construction apply equally to converted buildings.

Siting

Ideally the site should be reasonably level, unobstructed and capable of further development, with good access to public roads. Electricity and water supplies should be available, and there should be scope for the disposal of any waste products. The store should be in a central position and be integrated with activities such as grading, packing and processing, and preferably near associated housing, both to facilitate good management and discourage vandalism.

Whether or not planning consent will be necessary will depend on current legislation. Access to highways, heights of buildings, distances from public roads, choice and use of materials and general amenity aspects must be taken into account. Compliance with building and other regulations must be ensured and approvals obtained at the appropriate times. Professional advice is essential on all these aspects.

Types of stores

Refrigerated stores may be considered to be of two types: cool stores or freezers. The designer should be aware of the importance of the quick removal of field heat from certain crops prior to storage or despatch, and of vacuum and icebank cooling systems and refrigerated transport facilities. Some knowledge of current sale and display methods in shops and supermarkets can be useful but these wider aspects are not covered in this book.

Cool stores

These are stores, provided with refrigerating equipment, which are designed to operate at temperatures above or in the region of 0°C. They can be subdivided as follows:

Controlled atmosphere stores (CA stores), where the concentrations of oxygen, carbon dioxide and nitrogen in the store atmosphere may be varied from those in the external air.

Stores of similar design but where the facilities for atmosphere control are not provided.

Stores where refrigeration equipment is provided but is ancillary to ventilation with ambient air.

Jacketed stores where the produce-holding areas of stores are separated physically from the cooling air to maintain high atmospheric humidity. Since these stores are more likely to be used for ornamental nursery stock, their design is not covered in this book.

Freezers

These are stores designed to operate at temperatures appreciably below 0°C. Such temperatures are not usually required in horticulture but there may be occasions when stores of this type are necessary. Since problems peculiar to these stores tend to arise in their design, specialist advice is needed on their construction and they are therefore outside the scope of this book.

PLATE I
Apple showing superficial scald

PLATE II
Apple showing bitter pit

PLATE III
Apple showing physiological senescent
breakdown

PLATE IV
Apple showing brown rot

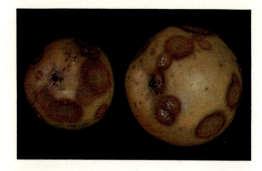

PLATE V
Apple showing *Gloesporium* rot

PLATE VI
Apple showing brown heart

PLATE VII
Apple showing *Phytophthora syringae*

PLATE VIII
Pear showing *Botrytis* fruit rot

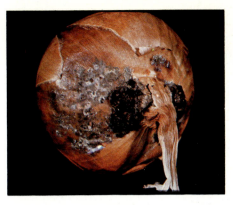

PLATE IX
Onion showing neckrot

PLATE X
Carrots showing licorice rot

PLATE XI
Carrots showing crater rot

PLATE XII
Red beet showing dry rot

PLATE XIII
Celery showing *Centrospora* rot

PLATE XIV
Cabbage showing *Phytophthora
porri* (Photo: FRI)

PLATE XV
Cabbage showing pepper spot
(Photo: NVRS)

PLATE XVI
Cabbage showing internal
necrosis caused by turnip
mosaic virus (Photo: NVRS)

PLATE XVII
Cabbage showing turnip mosaic
virus on outer leaves
(Photo: NVRS)

Typical layouts

Some methods of building and insulating stores are illustrated in Figures 4–9, and the explanatory notes indicate conditions which may be required. These constructional methods should not be regarded as the only ones available; provided recommended principles are followed, suitable alternatives can be adopted. Further details of Recommended U Values are given in Appendix 2.

Fig. 4   Store A is one of a free standing block associated with a fruit packing shed. Its purpose is to store apples at a temperature of 3·3°C during the period October–March. It is gas sealed and provides fully controlled temperature conditions

Fig. 5   Store B is one of a block of prefabricated, sectional stores, built within a wide span building. Its purpose is to store top fruit and it is suitable also for many vegetable crops. It is designed to operate at 0°C throughout the year or at minus 1°C for pear storage

Fig. 6   Store C is built within an existing packing shed. Its purpose is to hold horticultural produce for short periods; weekends, overnight and intermittently throughout the year. It is capable of operating at 2°C

Fig. 7   Store D is a free standing chamber formed within a portal building. It is designed to hold a variety of vegetable crops in boxes. Its ventilation system is based on ambient air, with refrigeration facilities

Fig. 8   Store E is one of three within a clear span building. It is designed as a versatile store for crops in bulk or boxes. Its operating temperature can be 0°C throughout the year, but it could be adapted for other uses by the addition of heating elements or, with design modifications, could be used for lower temperature storage

Fig. 9   Store F has an above ground main ventilating duct. It is designed to hold crops in bulk and refrigeration equipment can be brought into use if and when required

Construction of stores

MATERIALS AND WORKMANSHIP

Whatever the form of construction, all materials must be of good quality. Using inferior materials is false economy in view of the high cost of equipment and the value of crops stored. Materials should conform to British Standards, where applicable, and workmanship to the appropriate Codes of Practice. The importance of good workmanship cannot be overestimated, since mistakes are costly and difficult, if not impossible, to rectify.

Timber must be kept dry before use and protected from moisture when built into the structure. All timber should be treated under pressure or by the vacuum impregnation process. Chemicals used in the treatment must not be harmful to stored products or incompatible with adjacent materials. For example, creosote is clearly unsuitable because of the risk of tainting and aluminium should be protected from certain timber preservatives. For detailed advice reference should be made to Leaflet No. *17 Preservation of Timber and Metal* in the MAFF series of Fixed Equipment on the Farm Leaflets, available from MAFF (Publications), Tolcarne Drive, Pinner, Middlesex HA5 2DT.

Metals can be protected from corrosion with bitumen based paints; where different metals are in contact, such protection is essential. Adhesives and bituminous materials must be compatible with the surfaces to which they are applied:

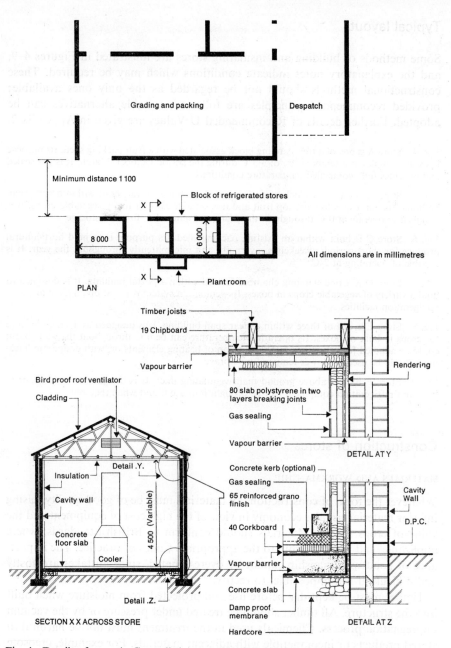

Grading and packing

Despatch

Minimum distance 1 100

Block of refrigerated stores

X

8 000

6 000

All dimensions are in millimetres

X

Plant room

PLAN

Timber joists

19 Chipboard

Vapour barrier

Rendering

Bird proof roof ventilator

Cladding

80 slab polystyrene in two layers breaking joints

Gas sealing

Vapour barrier

DETAIL AT Y

Detail .Y.

Insulation

Cavity wall

Concrete floor slab

Cooler

4 500 (Variable)

Concrete kerb (optional)

Gas sealing

65 reinforced grano finish

Cavity Wall

D.P.C.

40 Corkboard

Vapour barrier

Concrete slab

Detail .Z.

Damp proof membrane

SECTION X X ACROSS STORE

Hardcore

DETAIL AT Z

Fig. 4   Details of store A. Controlled atmosphere fruit store

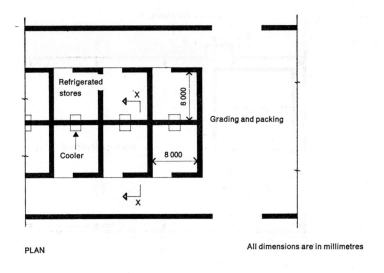

Refrigerated stores

Grading and packing

8 000

8 000

Cooler

PLAN

All dimensions are in millimetres

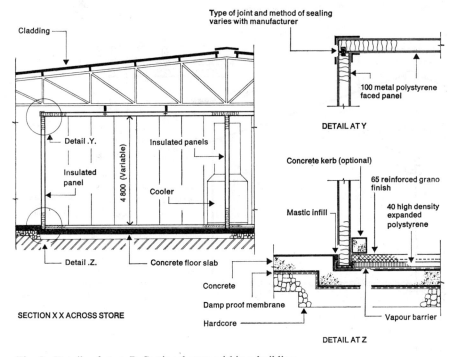

Cladding

Type of joint and method of sealing varies with manufacturer

100 metal polystyrene faced panel

DETAIL AT Y

Detail .Y.

Insulated panels

Insulated panel

4 800 (Variable)

Cooler

Detail .Z.

Concrete floor slab

SECTION X X ACROSS STORE

Concrete kerb (optional)

65 reinforced grano finish

40 high density expanded polystyrene

Mastic infill

Concrete

Damp proof membrane

Hardcore

Vapour barrier

DETAIL AT Z

Fig. 5   Details of store B. Sectional store within a building

69

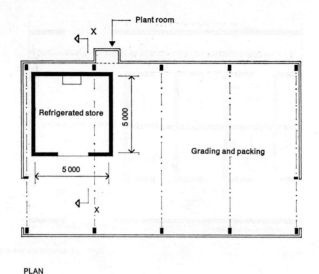

Plant room

X

Refrigerated store

5 000

5 000

Grading and packing

PLAN

All dimensions are in millimetres

Timber Framing

19 Chipboard

Existing building

Applied mastic finish

DETAIL AT Y

Continuous vapour barrier

80 slab polystyrene
in two layers
breaking joints

Timber framing to form store

Detail .Y.

Asbestos cement
sheet lining

Chipboard or
other suitable
external lining

3 000

Timber plate bedded
on D.P.C.

Detail .Z.

Existing concrete floor

450

SECTION X X ACROSS STORE

50 vapour sealed expanded plastic
perimeter insulation

DETAIL AT Z

Fig. 6   Details of store C. Short term store for horticultural products

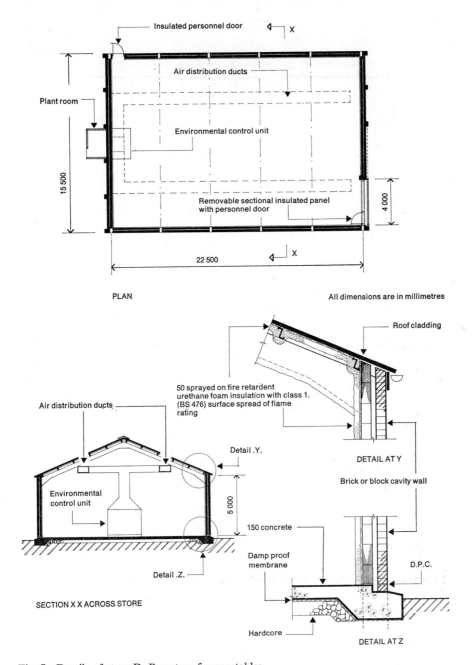

Insulated personnel door

Air distribution ducts

Plant room

Environmental control unit

15 500

Removable sectional insulated panel
with personnel door

4 000

22 500

PLAN

All dimensions are in millimetres

Roof cladding

50 sprayed on fire retardent
urethane foam insulation with class 1.
(BS 476) surface spread of flame
rating

Air distribution ducts

Detail .Y.

DETAIL AT Y

Brick or block cavity wall

Environmental
control unit

5 000

150 concrete

Damp proof
membrane

D.P.C.

Detail .Z.

SECTION X X ACROSS STORE

Hardcore

DETAIL AT Z

Fig. 7   Details of store D. Box store for vegetables

71

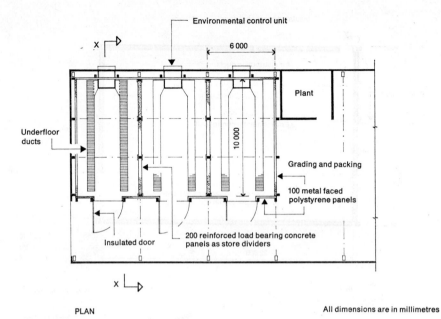

Environmental control unit

x

6 000

Plant

Underfloor
ducts

10 000

Grading and packing

100 metal faced
polystyrene panels

Insulated door

200 reinforced load bearing concrete
panels as store dividers

x

PLAN

All dimensions are in millimetres

Type of joint and method of sealing
varies with manufacturers

Steel support members

Protected insulation over steel

Vents with insulated covers

100 polystyrene panels faced both
sides with sheet metal

5 000

65 reinforced grano finish

40 corkboard

Vapour barrier

40 corkboard insulation to ducts

Environmental control unit

150 concrete

Damp proof membrane

SECTION X X ACROSS STORE

Hardcore

DETAILS AT FLOOR AND CEILING

Fig. 8   Details of store E. Box or Bulk store for vegetables

72

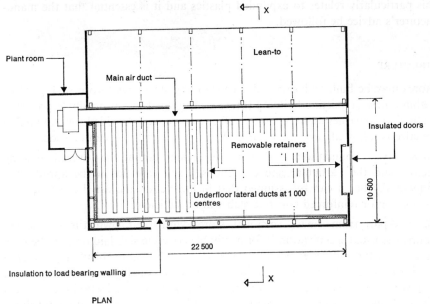

Plant room

Main air duct

Lean-to

Insulated doors

Removable retainers

Underfloor lateral ducts at 1 000 centres

10 500

22 500

Insulation to load bearing walling

PLAN

All dimensions are in millimetres

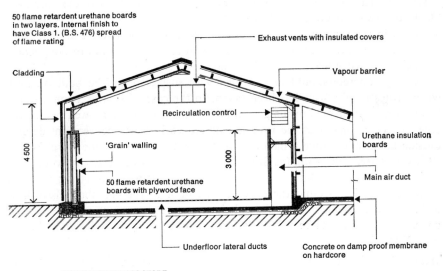

50 flame retardent urethane boards in two layers. Internal finish to have Class 1. (B.S. 476) spread of flame rating

Cladding

Exhaust vents with insulated covers

Vapour barrier

Recirculation control

Urethane insulation boards

4 500

'Grain' walling

3 000

Main air duct

50 flame retardent urethane boards with plywood face

Underfloor lateral ducts

Concrete on damp proof membrane on hardcore

SECTION X X ACROSS STORE

Fig. 9   Details of store F. Bulk store for vegetables

73

this particularly relates to expanded plastics and it is essential that the manufacturer's advice be followed.

STRUCTURE

Stores may be built as individual, free-standing structures (Store A, Fig. 4) or within a new or existing building (Store C, Fig. 6). There is no single preferred method of construction to be used in all circumstances. Provided sound building practice is followed, any one of several recognised methods can be used.

Foundations Foundation design is dependent upon the overall loading factors and the particular site conditions. Each case has to be approached individually but it can be assumed that in most cases the usual types of strip foundation or reinforced concrete slab will be sufficient.

Floor Apart from small, sectional stores with their own timber framed floors, the usual specification is for a suitable concrete slab laid on polythene or other sheet type of water barrier over a layer of hardcore. Insulating material will in many cases be incorporated in the floor. Good quality concrete may suffice as a wearing surface but, where fork lift trucks or other small wheel transporters are used, such a surface is liable to wear and deteriorate and a different finish will be required (see Internal Finishes, Doors and Hatches, page 79).

When under-floor ducts are necessary they can be built in bricks, concrete blocks or *in situ* concrete but they must not admit water. They may need to be insulated in the same way as the floors, depending upon overall design. Duct covers may be in timber, steel or concrete; loose brick floors are not recommended for refrigerated storage chambers.

Existing concrete should be regarded with suspicion and be built on only when professional advice has been taken. In the case of small sectional stores incorporating their own insulated floors, existing concrete may well be acceptable to receive them but the resulting difference in levels between inside and outside floors may prove a handicap to filling and emptying the store. Where such stores are built *in situ* (as Store C, Fig. 6), it may be necessary to provide perimeter insulation which consists of a vertical strip of insulating material placed in position around the edges of the floor of the chamber where heat leakage is usually maximum. This obviates the need to break up the existing concrete floor as would be necessary if a horizontal layer of insulation had to be incorporated.

Walls A distinction must be made between stores where the produce is held in containers and those where it is in bulk with the produce likely to exert a lateral pressure on the walls.

In the first case traditional methods such as solid or cavity brick or block work are satisfactory. The interior of the store must be kept dry and a horizontal damp proof course (DPC) included. (Store A, Fig. 4.) Alternatively timber framing may be used to build the store (Store C, Fig. 6) or the roof may be carried on stanchions with infilling of various materials. These include composite panels of steel, asbestos, aluminium or timber combined with insulating materials. Solid insulating concrete panels may be used between stanchions, and when suitably reinforced can also be used in bulk stores.

In the second case when crops are stored in bulk the walling must be designed so as to accept the lateral loads and transfer them safely to the foundations. The magnitude of these loads will vary with the crop and the height of storage and for guidance on design methods to be followed reference should be made to BS 5502 (Code of practice for design of buildings and structures for agriculture) and its loading Addendum. Methods of construction include reinforced brickwork or concrete or, more usually, some form of prefabricated panel. Grain walling systems are adequate structurally but may need modifications to make them suitable for vegetable storage (Store F, Fig. 9).

Where insulating panels are used as complete walling (see Store B, Fig. 5) precautions are essential to obtain sealed, tight fitting joints and continuity throughout. Where a store is being built within an existing building it may be possible to use the cladding to keep out the weather and to construct store walls entirely within this shell.

Roof Sound construction and good workmanship are particularly important. In some cases a flat roof may be unavoidable but wherever possible it is preferable to construct stores with a pitched roof. Where a flat ceiling is required within the store, the space above this and below the pitched roof can be used for inspection and access to both stores and fans, and defects in the construction of the roof and ceiling can be seen and put right without delay.

Most pitched roofs are built with steel, timber or concrete frames with purlins carrying sheet materials such as steel, asbestos or aluminium. If constructed to good practice, there should be few problems but it is essential that laps and joints at ridges, verges and eaves be sealed and bedded closely to prevent fine snow entering. Although ventilation should be provided, roof spaces should be kept as dark as possible inside to discourage entry of birds. Other roof surfaces employing flexible sheet materials may be used, in accordance with sound, current building practice.

Ceilings Where a flat ceiling is required, as in most CA stores (see Store A, Fig. 4), it is usually formed from timber joists spanning between chamber walls, trusses or special steel and timber beams. In every case it is essential to design the framework so as to adequately accommodate the loads to be applied and to ensure a rigid framework with the minimum of deflection. Suitable materials for lining and insulation are discussed later on page 76.

In some cases supports for the walls and ceilings can be taken from the roof members over the store, using steel hangers and runners. In the case of sectional stores erected under existing roofs, the ceiling panels must be designed as self-supporting units or suspended from external frameworks or from the main roof in preference to relying on support from beams within the store itself (see Store B, Fig. 5).

Sectional stores Recent developments in the design of sectional stores for boxed produce call for special consideration (see Store B, Fig. 5). Prefabrication of the various parts of a store, followed by assembly on site is now a recognised alternative to traditional methods of construction. In order to reduce the site work even further, complete designs of sectional stores are offered as package deals. Such stores may be supplied complete with refrigerating equipment and controls. The use of pre-formed, insulated, structural panels has made it possible to reduce to a minimum any ancillary supporting structure—with obvious advantages. It is sometimes difficult to obtain details about the design of the panels and the method of fitting them together, but in all cases the principles of stability and suitability for performance must be followed and the proposals scrutinised accordingly.

Some proprietary systems use timber in the construction of panels in walls and ceilings and this timber must be pressure treated with preservative before fabrication. Ceiling and roof support systems should be examined, also the methods of erection and fixing. Particular points to note are the panel junctions and methods of sealing; structural support for ceilings and general structural qualities. The methods used for supporting door hangings and hinges and the avoidance of unnecessary bolt and screw holes through the panels are other important items. The intending purchaser should always inspect examples of proprietary stores already in use before deciding to purchase stores of similar manufacture.

INSULATION

To limit the load on the refrigerating equipment, heat leakage into a store must be reduced to as low a level as is practicable. Insulating materials are incorporated in the store for this purpose, the method of application varying with the materials selected. The most important characteristics of such materials are: resistance to moisture transmission, ease of installation, absence of taint and, where exposed, low surface spread of flame.

Suitable materials can be found within the whole range of commercial, thermal insulants but in practice they tend to be selected from certain well-known groups and include board or slab materials of a rigid or semi-rigid character, granular or loose fill materials, materials foamed *in situ* or sprayed on insulants. Some insulating materials, in particular plastics of various kinds, can constitute a fire hazard, both from combustibility and spread of flame, and when burning, from the production of smoke and toxic fumes. Professional advice should be taken where there is any doubt about the characteristics and safe use of a material.

Examples of some types of insulating materials for refrigerated stores are given below, but this list is not complete. The Table in Appendix 2 lists some of the more usual materials and indicates their effectiveness as shown by their k value (the k value is a measure of the heat transmission qualities of the material; the lower the k value the better the insulating value of the material). From the k values of the various materials used in the structure of a store, the overall insulating value of that structure can be derived.

Board or slab insulations These include plastics such as expanded polystyrenes and polyurethanes which, together with cork slab, are the most commonly used materials for lining the inside of store walls, ceilings and roofs. Plastics have tended to displace the more traditional cork due to the advantages of price, availability and weight. Methods of fixing include nailing, skewering with hardwood pegs and adhesives. As with cork slab, plastic materials can be used horizontally below the floor finish but high density quality must be selected to resist the concentrated loading which may arise when produce is being

76

mechanically handled into and out of store. Slab insulations are also suitable for use in composite panel construction and for filling cavities.

Granular and loose fill materials   These include regranulated cork, mineral wool and pelleted polystyrene. It was formerly standard practice in some fruit store designs to use regranulated cork, either between brick skins or within cavities in stud walls lined with timber or galvanised steel. Where a good vapour barrier was provided and where the material was kept dry, the result was satisfactory but all granular materials are liable to settlement and it is almost impossible to check this settlement in vertical cavities once a store is completed and in use.

Foamed 'in situ'   Chemical foams and other insulating materials can be introduced into cavities in a similar way to granular materials. They can be effective and in certain circumstances the method is useful. It is important not to overlook the fact that large quantities of water are produced in the process of foaming and this water must be allowed to dry out of the structure.

Sprayed-on insulation   Urethane foams sprayed onto walls, roofs and ceilings can be satisfactory but the application is a specialised one and particular care is needed in the choice of protective finishes.

Pre-formed panels   Factory produced panels utilising one or more of the above insulants are increasingly used and are of many types. Galvanised steel and aluminium sheets, rigid plastics, plywood and flat asbestos are all available as the exterior components of the panels. Plastic insulating materials have several advantages in forming these panels, including relatively low cost, lightness and ease of use and manufacture. Incorporating the materials within the panels protects against mechanical damage to which they are prone; in this situation they are also a reduced fire risk and are less likely to be damaged by vermin.

Lightweight concrete   Insulating panels and blocks are made from selected aggregates and cement. They are light and the panels can be reinforced where necessary to take lateral loads. They are liable to damage from impact but can be rendered for protection.

INSULATION THICKNESS

To determine the thickness of insulation required in any particular case, it is necessary to decide upon a theoretical U value for each element in the construction, that is, for the wall, ceiling and floor. (The U value is a measure of the heat transmission characteristics of the structure, the lower the U value the better the insulation value. See Appendix 2.)

Calculations are necessary involving the store temperatures required, the periods of use, the type of equipment selected, the capital and running cost relationship, the degree of exposure of the store to prevailing winds, its actual construction, and the acceptable loss of weight from the produce. For example, if low temperatures are required during the summer months, running costs can be high unless an appropriate thickness of insulation has been provided. Also, if defective construction leads to unnecessary air changes in the store from outside wind movement, then the theoretical values of the insulation will not be achieved. Thus each store has to be examined and calculated separately. There are, however, some broadly accepted figures for U values for particular uses and those recommended for the stores illustrated are given in Appendix 2.

Whatever the insulation, to be fully effective it must be kept dry and protected from any mechanical damage that might occur after installation. Moisture can reach the insulation in stores from several sources including: rain and snow, rising damp, from defrosting cycles, or from water present in the constructional materials during the course of erection of the store. Dampness from these sources can be prevented by proper design but a further problem may arise from the migration of water vapour from the outside air into the store via the insulation itself. Since there will normally be a temperature difference between the store air and the outside air there will be a movement of water vapour from the warmer to the colder side of the insulation and the greater the temperature difference between the inside and outside air, the greater will be the movement of the water vapour. There will be a tendency for the moisture to be extracted by the refrigerating plant but every effort must be made to reduce to a minimum the amount entering the store in order both to improve the efficient running of the store and plant and to minimise the risk of dampness in the store insulation. This is done by the provision of a vapour barrier on the warmer side of the insulation.

In some forms of construction as for instance where metal clad panels are used, the cladding itself is an effective barrier but at joints, angles and junctions care is needed to ensure continuity by cover strips or special tapes. Materials suitable for use as vapour barriers include metal foils, polythene sheet and bitumen based compositions. Where the insulation is to be applied after erection of the frame of the store, it is usual to provide a level and true surface to walls and ceiling on which the vapour barrier may be applied in a complete and continuous form to join up with a horizontal barrier of the same or other suitable material within the floor construction. The insulating material can then be applied to the interior. It is essential that all vapour barriers be continuous and unbroken throughout and this can only be achieved if there is careful control over all stages of design and construction. Some suitable materials for vapour barriers are listed in Appendix 2 together with an indication of effectiveness.

GAS SEALING AND SCRUBBING

Where a CA store is required vapour barriers alone are insufficient for gas tightness. A store must conform to strict test conditions and be held to a specified rate of pressure drop for a definite time. This test cannot be successfully achieved without particular care being taken, both in design and construction and in the selection of gas-proof lining materials. The danger of air leakage into a store through the construction itself has to be constantly borne in mind and all doors, hatches and access points must be close-fitting with built-in gas seals.

It has been customary to use one of two methods to gas seal farm type stores. Ceilings, walls, doors and hatches can be lined with galvanised flat steel sheeting, greased back and front with petroleum jelly and bedded onto timber frames. The sheets are overlapped at least 50 mm using heavier grease at the laps and nailed frequently with galvanised, flat headed nails. Doors are finished in the same way and, in addition, have double rubber seals and, at the floor, removable thresholds bedded in grease. In many cases additional metal gas sealing panels, bedded in grease, are used. When carefully installed the method is effective. A gas sealing compound is painted on the floor surface or applied integrally with the finish itself; this is of particular importance in new stores operating at high carbon dioxide levels.

The normal alternative to galvanised sheet lining is to apply by trowel a bituminous gas proofing compound over the entire internal wall and ceiling surfaces. Doors, hatches, openings and floors still require treatment as already described. This method needs attention each year to ensure that cracks are filled and damaged areas repaired.

There is a proprietary method in which a gas sealing coating based on a rubber derivative is applied over a glass fibre strengthened base coat. Provided that the backing is of a suitable nature, this appears effective in producing a clean, light surface which does not require further treatment. Repair is simple in the event of damage but care is required in all stages of the application particularly where metal sheeting is involved.

Where metal faced panels are used in a sectional construction, it is possible to seal the joints between them and so obtain a fully CA store. The floor, door and other access points need special treatment as described above.

Scrubbing If lime is to be used to remove carbon dioxide from the store atmosphere, it is usually necessary to build a separate room or cabinet to hold the lime and this has to be connected to the store. It must be constructed to the same standards of insulation and sealing as the store itself. Other systems for scrubbing may need space in the plant room or elsewhere.

Special precautions In addition to the careful attention which must be given to each detail of construction, it is essential to make sure that the vapour barrier is undamaged and fully effective. If moisture vapour is allowed to migrate from the outside through the insulation it will be trapped by the gas seal on the inside of the store provided specially to give CA conditions with little opportunity of drying out (see Internal Finishes below). The use of sealed, insulated panels which guarantee that the insulant will remain dry under all conditions may offer the most satisfactory means of ensuring maximum efficiency in CA stores. There remains then only the problem of providing a gas proof junction to the panels and to any openings within the store. Provided there is rigidity in the store construction, this problem is not insuperable using available plastic materials.

INTERNAL FINISHES, DOORS AND HATCHES

Walls and ceilings Any internal finish applied should theoretically be of a less permeable nature than the vapour barrier on the warm side of the insulation to avoid trapping moisture as already discussed. This problem does not arise with fully sealed sections, apart from the joints; the internal face can therefore be treated in a variety of finishes or pretreated with a factory applied finish.

In some cases it may be practicable to leave the internal surfaces without further treatment, provided there is no fire risk. Insulating blocks or panels may be rendered or left untreated; masonry generally can be rendered, as can cork slab and polystyrene. A flexible type of plaster can be used on walls and ceilings, and decorative treatments of many sorts are available. Such treatments can be arranged to reduced fire risks, inhibit mould growth, and to provide a light, easily cleaned interior. They should be selected from the range recommended by manufacturers as suitable for storage work.

Interior surfaces should be lined with metal, asbestos, plywood or hardboard only when the principles above are understood and followed. The fixing of such linings, as in the case of gas sealing with metal, requires the provision of timber grounds (unless the materials are pre-bonded).

Floors Where a finish has to be laid on top of a layer of insulation, 65 mm of granolithic concrete is normally satisfactory. To avoid cracking it is good practice to reinforce this layer with light steel fabric. Where exceptional conditions involving heavy wear, particularly point loads from small, steel wheeled traffic are encountered there are a number of proprietary, hard wearing, industrial type floor finishes available. The advice of manufacturers and suppliers should be sought in particular cases to find the most suitable to meet the particular store conditions.

Doors Door openings must be large enough to avoid damage to the store structure at times of filling and emptying but, because they can be a major source of heat entry, they should not be made larger than necessary. Dimensions will depend on store size, the means used for loading and unloading, and the type of container. The minimum width should be twice that of the largest unit likely to pass through, and in the case of bulk stores may have to be much greater than this. Where industrial type fork lift trucks are used to stack pallets, openings will usually be about 4 m wide. Height must be such that a safe clearance is retained above loads, and in some circumstances openings to the full height of the chamber are necessary to facilitate loading and unloading. Apart from bulk stores, an entry door is best sited at one side of the entry wall, rather than in the centre, since this usually permits easier loading.

Doors must be insulated to the same degree as the store itself, must contain a vapour barrier and have an internal finish equivalent to that of the store walls. The design and construction of doors calls for special skills and in consequence they are usually purchased as a factory made article from insulation contractors to the performance specification required. If timber is included it must be pressure treated with preservative before fabrication. Expanded plastic or polyurethane and claddings of sheet metal, either painted or with a factory applied finish are the usual materials. Doors may be hinged or sliding. In the case of large, hinged doors a support wheel on the outer edge can help to reduce the load on the hinges. In all cases, frames must be completely rigid and both doors and frames entirely free from twisting or warping. All fixings and fastenings must be robust and permanent with a positive action. It should be possible to adjust the tension of the locking devices to ensure equal and satisfactory closure against the synthetic rubber gaskets. These are set in rebates in the frame or, in the case of sliding doors, fixed to act directly against the face of the door. It is possible to obtain a positive sealing action by using a proprietary design in which the door drops slightly into its final position. In all cases, the design of the threshold is important, since it must not obstruct the free entry of wheeled traffic. Various designs are in use; a common one is the removable threshold which can be secured in position before the doors are closed. More specialised systems are designed by the manufacturers of sliding doors. In all cases an alarm system operated from inside the store must be provided or an internally operated method installed by which the doors can be opened should anyone be trapped inside.

Hatches   Hatches are usually provided to give access for inspection from the ceiling. They should be at least 750 mm square and, where necessary, large enough to allow the removal of a fan: their design is important and must match the rest of the store in insulation value and general construction. Suitable locking devices must be provided, together with seals as in the case of the doors.

Engineers' plates   All service pipes and cables entering the store should be grouped and brought into the store at one pre-determined point. A suitable entry plate should be provided through which the services can be passed and should be built in during the erection of the store. It should be insulated on the outside to avoid condensation.

Condensate outlet   Beneath or adjacent to the cooler, a drain outlet must be provided to take water condensing on the coils and collecting as a result of defrosting operations. This drain outlet should be fitted with a trap 100–150 mm deep and lead to a suitable position outside.

Entrance precautions   Where there is much traffic into or out of a refrigerated store, considerable loss of efficiency can occur due to the entry of warm air. Air locks or secondary, flexible doors can be provided but will tend to interrupt free movement. Where there is a real demand air curtains can be provided but these are expensive. Several manufacturers offer equipment for this purpose which it is claimed reduce air change within the chamber to a minimum, and they are prepared to assist designers.

Electrical services

All electrical installations must conform to the latest issue of the *Regulations for the Electrical Equipment of Buildings* published by the Institution of Electrical Engineers. This includes a special section covering the problems associated with agricultural and horticultural installations. In the case of cool stores, high humidity conditions and the resulting problems of corrosion call for special care. Several wiring systems are acceptable, the choice depending upon the special requirements of each individual case.

An all-insulated wiring system is suitable where wiring and accessories can be positioned to avoid impact damage but where additional protection is needed, heavy duty plastic conduit is required. In the case of wiring to motors or in any other exceptional circumstances, heavy gauge galvanised screwed conduit may be essential. Other types of conduit are not acceptable. Entry points for cables into stores must be via specially designed sleeves or the engineers' sealed entry plate with an insulating mastic material to seal off the outer and inner atmospheres. Connections at switches, junction boxes and light fittings must be filled with suitable, waterproof, mastic material and the light fittings within the store must be of a waterproof type. The light intensity provided needs to be equivalent to that in a normal packing shed to ensure quality control. Switches must be located outside the store and clearly marked *On* and *Off*. Fixing 'grounds' for conduits, wiring systems and fittings must be provided during the construction of the store. Wiring must be on the surface and not buried in walls, floor or ceiling.

Similar conditions as to lighting apply at the plant room and round the exterior of the store where produce is being handled. Enough socket outlets need to be provided outside the stores and located in suitable positions. Switch gear control panels and instrument controls should be located centrally and in a position that will ensure easy access at all times. They must be protected from unauthorised use and damage. All installation work must be carried out by qualified and experienced contractors. Layout diagrams should be provided and kept for reference in a safe place easily accessible to staff. In addition to the completion certificate that must be provided as described in the IEE Regulations, periodic tests must be carried out to ensure that the installation is maintained in a safe condition. In cases of doubt advice can be obtained from the local offices of the Electricity Board.

Vermin proofing

It is impracticable to design buildings which are completely vermin proof but precautions should be taken against damage by rats, mice and birds as part of a comprehensive pest control policy on the holding. Advice is available from the Ministry's pest control officer located at the local Divisional Office. The following Ministry publication also contains recommendations for the proofing of buildings against entry of rats and mice, GBI (formerly Technical Bulletin 12): *Proofing of Buildings against Rats and Mice* price £1·25 plus postage available from HMSO or any bookseller.

To minimise pest infestation all holes in the store fabric should be filled in, cavities should be sealed and ducts and outlets protected. Any roof space with exposed plastic insulation must also be protected. To deter rats and mice at ground level 300 mm high, flat, galvanised steel sheets can be fixed at the base of wooden doors or similar vulnerable situations. The area round the store must be kept clean and free from rubbish, and any possible nesting sites cleared away.

Safety

All buildings, machinery and other equipment must comply with the requirements of the Health and Safety at Work, etc. Act 1974 and the Agriculture Safety Regulations. Readers should be satisfied that these requirements will be met before finalising proposals and if in any doubt should consult the Agricultural Inspectorate at the local office of the Health and Safety Executive.

Note: Since this book was written, BS 5502 Code of practice for Design of building and structures for agriculture has been produced and is obtainable from British Standards Institute, 2 Park Street, London W1A 2BS. Where recommendations conflict, those given in BS 5502 should be followed.

Plant and equipment

The following section describes the plant and equipment that may be used to provide the correct temperature, humidity and atmosphere for the storage of fruit or vegetables.

Equipment to control temperature

In the majority of stores store temperature will be controlled by means of a refrigeration system. Mechanical refrigeration is used almost exclusively in agriculture and horticulture. Other systems of refrigeration may be used in small, short-term stores, and for refrigerated transport. These include the use of liquid nitrogen and solid carbon dioxide.

THE BASIC REFRIGERATION CYCLE

The basic, vapour compression, refrigeration system consists of a closed pipe circuit connecting four essential components; the compressor, the condenser, the expansion or regulator valve and the evaporator. The *compressor* raises the pressure of the refrigerant vapour and delivers it to the condenser. In the *condenser* the vapour cools and is condensed into a liquid. During the change of state from vapour to liquid heat is given up. The liquid refrigerant is then passed through the *regulator valve* into the *evaporator* which is at a lower pressure. The liquid refrigerant evaporates at a lower temperature under the reduced pressure in the evaporator. The necessary heat to convert the refrigerant from liquid to vapour is drawn from the materials surrounding the evaporator. The vapour is returned from the evaporator to the compressor to be re-used.

Direct expansion

Stores in which the *evaporator* is located inside the store are referred to as 'direct expansion systems'. This arrangement is very widely used for small and medium sized stores because it is simple and the least expensive.

Normally a separate *condensing unit* is provided for each store. In large installations it is possible to arrange for two or three compressors to supply liquid refrigerant to all the storage chambers. Such an arrangement allows flexibility in store management since cooling capacity can be directed to where it is needed. Further advantages of such an arrangement are that the condensing capacity can be matched more exactly to the cooling requirement of the stores

83

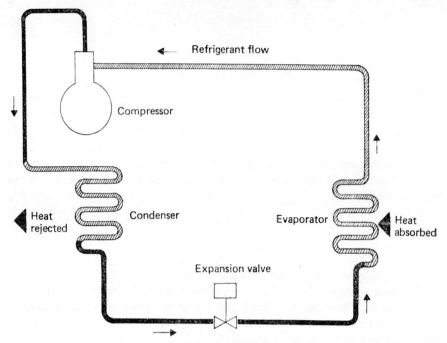

Fig. 10   Vapour compression refrigeration cycle

over a wide range of conditions. Where high humidity is important it is advantageous to be able to reduce the plant capacity by switching off one or more compressors once the produce has reached storage temperature. The failure of one compressor need not terminate the storage of produce, because one of the other units can act as a standby whilst repairs are carried out.

Indirect expansion

In this type of plant the evaporator is immersed in a secondary cooling medium (coolant) which is pump circulated through the coolers within the stores. The flow of coolant to each cooler is automatically controlled by a valve. Indirect cooling is only considered economic on large installations such as fruit or vegetable co-operatives where a large number of stores are run at similar temperatures. The secondary coolants used in indirect systems (calcium chloride brine or ethylene glycol) are corrosive. The corrosion of valves, coolers and pipe work can be a problem so care is needed in the selection of materials for these components. Indirect systems are more expensive to run than direct ones because of the heat gains from coolant circulating pumps and pipe work.

Well designed indirect systems can provide more accurate temperature control and higher relative humidities than simple direct expansion plants. The indirect system also gives maximum flexibility since cooling can be directed to where it is needed. In the design of an indirect system it is desirable to install two compressor units each supplying half the maximum cooling duty. If this is done, one unit can act as a standby for the other in case of one unit failing. Such an arrangement also leads to a better match between cooling requirement and refrigeration capacity once the produce has been cooled. It is also considered good practice to duplicate coolant circulating pumps.

Ice bank coolers

In installations where rapid cooling is important, it is necessary to install conventional refrigeration equipment which has sufficient capacity to deal with the peak cooling requirement. Once the produce has been cooled a large proportion of the installed plant capacity is no longer required: typically the peak cooling requirement is six times that required to keep the produce cool. In order to avoid the provision of excess capacity it is necessary to be able to store the cooling effect produced by the refrigeration plant. Where store operating temperatures are close to 0°C an efficient way to store refrigerating effect is to create blocks of water ice.

DIAGRAM OF ICE BANK COOLER USING POSITIVE VENTILATION

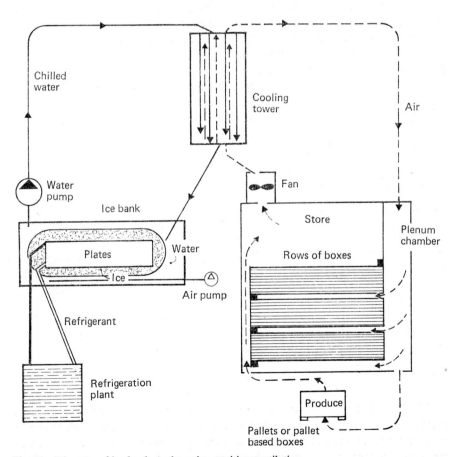

Fig. 11 Diagram of ice bank cooler using positive ventilation

The principle of the ice bank cooler is not new but its use has largely been restricted to the cooling of milk. The technique can be easily adapted to the cooling of vegetables.

The principle of operation is that a refrigeration compressor with sufficient capacity to meet the average daily cooling requirement is run continuously to produce blocks of ice. When it is required to cool produce, the cooling effect is released by melting the ice. Under suitable conditions the ice can be melted much more rapidly than it was made; thus enabling the cooling to be completed in a time only limited by the rate at which the produce can give up its heat. A further advantage of the ice bank system is that if correctly designed it will produce very high humidities in the store, minimising the weight loss that inevitably occurs during the cooling process.

Direct refrigeration

In some cases, where the annual use of facilities is very short, the installation of a vapour compression refrigeration system cannot be justified. In such situations it may be practicable to provide refrigeration by evaporating liquid nitrogen, or solid carbon dioxide. The high cost of the refrigerant and the fact that it is lost and cannot be recirculated, tends to exclude this method of refrigeration from applications which require the quick extraction of large quantities of field heat. It can, however, be successfully used for refrigerating transport vehicles to prevent cooled produce from warming up too quickly on its journey to market. It is important to remember that direct refrigeration systems of this type tend to influence the atmosphere composition. If care is not taken produce may be damaged, though in general terms the modifications in atmosphere may be slightly beneficial. Care must also be taken by personnel who enter the storage chamber. Adequate time should be allowed after the doors are opened to ensure that the atmosphere is normal.

Cooling with ambient air

For some crops the use of refrigerated storage facilities may not be justified. An improvement in storage life can be obtained by selective ventilation, using outside air. The principle of operation is that only air that is colder than the stack of produce is used for ventilation. It will be obvious that such a system depends for its effectiveness on an adequate supply of air at a temperature below the required storage temperature. As storage is extended into the spring, the number of hours available for ventilation decreases. In general, one may not with certainty maintain sufficiently low storage temperatures much beyond the beginning of April. For storage beyond this period it will be necessary to consider mechanical refrigeration, possibly as an adjunct to ambient air cooling.

QUICK COOLING METHODS

Perishable crops like soft fruit and salad vegetables benefit greatly from rapid removal of field heat. Several techniques are available by which produce can be rapidly cooled. Whichever method of cooling is adopted it is important to recognise that once the produce has been cooled it must be kept cool if the

86

potential benefits of cooling are to be fully realised. The high speed cooler must be backed up by cooled storage facilities.

Forced air cooling

The rate at which heat can be transferred from produce to an air stream depends firstly on the temperature difference between the crop and the cooling air, and secondly on the speed at which the air passes over the crop. The more intimate the contact between the air stream and the crop, the less important is the speed. From the foregoing comments it will be seen that the size and shape of container and the method of stacking within the cooler will have a considerable effect on the cooling times achieved.

It may, if the size of an installation warrants it, be worthwhile considering a separate forced air cooling chamber where produce can be reduced to within a few degrees of storage temperature in a fairly short time, subsequently being removed to a separate building for storage. Admittedly such a system may involve double handling but with careful design this could be reduced to a minimum.

FORCED AIR COOLING

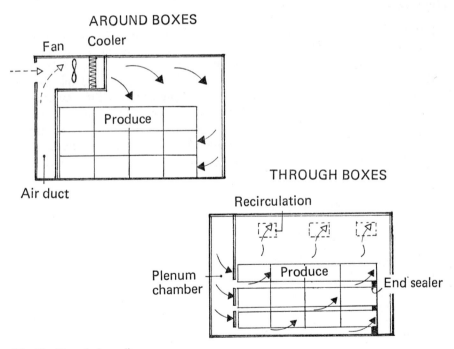

Fig. 12 Forced air cooling

Vacuum cooling

An alternative method for the rapid removal of field heat from produce is the vacuum cooler.

Most agricultural produce contains a plentiful supply of free water, so that when the produce is subjected to a suitable vacuum, some of this water is evaporated taking its heat of vaporisation from the produce and thereby cooling it. In fact, to cool produce by 10°C requires on average the evaporation of water equivalent to 1·8 per cent of the weight of the produce.

The mechanism of vacuum cooling is such that the cooling effect is generated within the produce. This has the important advantage of enabling even and rapid cooling of tightly stacked produce. It is even possible to cool produce that has been pre-packed ready for market, providing some ventilation holes are left in the packing materials. Vacuum cooling is most successful with flowers, mushrooms and with thin leafy produce, for example lettuce. It is perfectly possible to cool solid products like apples and strawberries but the larger the individual items are the longer it will be before their centre temperatures have been reduced to the required level.

By comparison with other cooling systems, vacuum cooling requires a higher capital investment.

Provided that the cooler is fully used for a long season the costs of cooling by this method are comparable to refrigerated cooling techniques. The vacuum cooler has the advantage of rapid uniform cooling regardless of method of stacking; an attribute which cannot be ascribed to forced air cooling systems.

Hydro cooling

Hydro cooling is an effective method of quickly removing field heat. Cooling times are comparable with those achieved by vacuum cooling. The method suffers from several disadvantages. The produce comes from the cooler in a very wet state which may make packing or handling difficult. The cooling water must be recirculated to retain the cooling effect and as a result it may become heavily contaminated with soil, plant sap, plant debris and disease organisms. Such conditions are conducive to the spread of fungal and bacterial disease which may in turn considerably shorten the shelf life of the produce. In some circumstances the disposal of used cooling water may present a serious effluent problem.

COMPONENTS OF REFRIGERATION SYSTEMS

Compressors and condensing sets

Positive displacement single stage compressors are used almost exclusively in agricultural and horticultural storage work. Direct drive by electric motor is most common, though some machines may be driven via a belt.

The compressor is often mounted in a frame which also carries the liquid receiver, refrigerant filter/drier, sight glass and the condensing coil together with its cooling fan. Such a unit is called a condensing set. When specifying the cooling capacity of a condensing set, it is essential to state the condensing and evaporating temperatures to which the capacity applies. The cooling power of a condensing set varies considerably at different temperatures, and for this reason the size of the compressor motor can only be considered as a guide to plant capacity. Condensing sets function most effectively when they are mounted in clean, cool positions. Although space and pipe work may be saved by installing

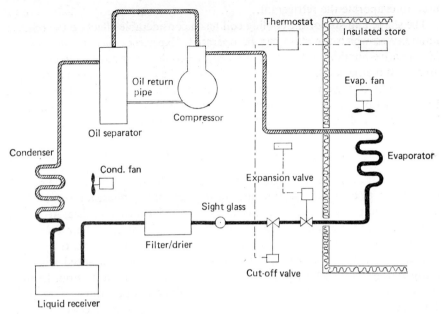

Fig. 13   Direct expansion cooling

the plant on the roof of the store, this is not recommended. Such a site is inaccessible for maintenance and cleaning. The air temperature is often well above ambient and vibration may damage gas barriers in controlled atmosphere stores.

Remote condensers

In some installations it may be necessary to locate the condenser away from the compressor. Two types of remote condenser are available. The air cooled condenser consists of a finned tube block, similar to a vehicle radiator, and a fan which blows air over it. This simple inexpensive arrangement is generally used on medium and small sized plants. The alternative, a water cooled condenser is more expensive, but is often preferred on larger plants because of the increased efficiency of the refrigeration system which results from the lower condensing temperature that can be obtained during warm weather. Water cooled condensers provide for greater flexibility in positioning since they do not have to be located near a good supply of cooling air. It is generally necessary to economise in the use of water; for this reason it will be desirable to recirculate the cooling water through a cooling tower. Water cooled condensers are generally only considered economic for plants rated at more than 50 kW refrigeration. With water cooled systems precautions must be taken to prevent possible frost damage.

Evaporator/Cooler

The majority of horticultural and agricultural stores are cooled by direct expansion. In such a system the evaporator performs the dual function of

89

evaporator and store cooler. Heat is extracted directly from the store air and used to evaporate the refrigerant.

The shape and size of the cooling coil has a considerable effect on the relative humidity at which the store can be operated. Experience has shown that if a high relative humidity is to be maintained, the cooler must have a larger surface area (8 sq metres/kW). This surface area should be arranged in such a way that the majority of the air passing through the cooler contacts the cooling surface. In order to provide a sufficiently large surface area within the confined space allowed for cooling coil, it is necessary to use finned tubes.

Air circulating fan

In order to transfer the heat from the produce in the store to the cooling coil it is necessary to circulate the store air. During the cooling phase a relatively high air circulation rate is required to ensure a rapid transfer of heat from the produce. Once the produce is cool, however, a much lower rate of air circulation is acceptable, and will result in much less weight loss from the crop. It is normal practice to arrange for the air circulating fan to run at half speed when the refrigeration compressor is off and to run at full speed when it is on. In some cases, provision is also made for reversing the direction of airflow in the store. This facility is claimed to be of benefit in minimising the effects of poor air distribution during storage. It is important, however, that the fan should be operated in its design direction during most of the storage period, since when it is running backwards, its efficiency is lower and there is an increased tendency for the cooling coil to become iced up.

DEFROSTING SYSTEMS

Under some operating conditions, for example, store temperatures below 3°C, and during the initial cooling period in most stores, the water vapour condensing on the cooler freezes to form a layer of ice crystals. The effect of this ice is to reduce the cooler's capacity to remove heat, and to increase its resistance to air flow. For efficient operation, it is therefore necessary to ensure that only a limited amount of ice is allowed to form. Three principal methods of defrosting can be used.

Air defrosting or off cycle defrosting

Air defrosting or off cycle defrosting is the simplest and cheapest method. It consists of allowing the store air to circulate over the cooler during the period when the refrigeration plant is inoperative. It is generally only suitable for applications where the store temperature is above 3°C.

Reverse cycle defrosting

In this system, the direction of circulation of the refrigerant is changed by means of a reversing valve. The functions of the condenser and evaporator are reversed, heat being drawn from the outside atmosphere and transferred into the store cooler. This system of defrosting is largely restricted to small direct expansion air cooled condensing sets. It may be particularly applicable where the maximum electrical loading at the site is limited.

Electric defrosting

This method involves the incorporation of electric heating elements in the cooling coil. It is suitable for all cooling systems.

Defrost controls

Defrosting is usually controlled by means of a time switch and an interval timer. The time switch initiates a defrost and the interval timer determines its length (usually 15–20 minutes). The number and length of defrosts given each day must be determined by experience. For well designed stores operating at 3°C and above, little or no defrosting should be required once the produce has reached storage temperature. Under such conditions the cooling coils will be defrosted between periods of plant operation by the circulation of the store atmosphere. For stores operating below 3°C, it will normally be necessary to provide a defrosting system. The frequency of defrosts in this case will depend on the type of crop. It is normal to expect more frequent defrosting to be required during the cooling period. In most cases some of the frosting that occurs during cooling may be avoided by initially setting the store thermostat 3 or 4°C above the desired storage temperature, then when the produce has been cooled to this temperature the thermostat is reset to the storage temperature.

Equipment to control humidity

The relative humidity of the store atmosphere is a measure of its drying power. The minimum drying power is associated with the maximum relative humidity. Since the majority of crops that are stored lose moisture and hence quality easily, it is important, particularly for long-term storage that the store humidity is maintained at as high a level as is practicable. If care is taken at the design stage, it should be possible to avoid the need for any kind of humidifying equipment within the store. At this stage, the choice of a suitable cooler, with adequate surface area and the correct evaporating temperature for the refrigerant, will go a long way to avoiding the need for humidification.

INCREASING HUMIDITY (HUMIDIFICATION)

The simplest method of providing humidification in a store is by spraying finely divided water into the air stream after the cooler. If this is done it is important to avoid liquid water being carried over into the produce, and wetting it, possibly resulting in an increased rate of rotting. When water is added to an already fairly moist storage atmosphere, little or none of the water may evaporate.

Where humidifying equipment has been installed to correct an original design error in the cooling system, care must be taken that sufficient defrosting time is allowed. The addition of water to the store atmosphere in such a situation is likely to result in quicker icing of the cooling coils, which will in turn lead to lower evaporating temperatures, and even more desiccation of the produce.

An alternative method, which has been adopted in some cases for improving humidity in existing stores, is to spray water over the cooling coil. This effectively increases the surface area available for heat transfer with the air and has the same effect as increasing the coil surface area at the design stage. Care must, however, be taken to ensure that water does not freeze on the cooling coil and cause it to become blocked.

Where it is available the use of a controlled quantity of steam is by far the most effective means of humidifying store atmospheres. The advantage of using steam lies in the fact that it requires no extra heat for evaporation since it is already a vapour. Careful control of the quantity of steam admitted is essential, since over supply would result in a considerable increase in the refrigeration load.

DECREASING RELATIVE HUMIDITY (DEHUMIDIFICATION)

For the effective long-term storage of, for example, onions, it may be necessary in addition to cooling the crop to remove moisture from it. The cooling equipment is capable of performing this function if suitably controlled. Where a store will be used exclusively for the storage of products requiring low humidity, the selection of a suitable cooling coil to provide the required conditions is an acceptable solution. In many cases, however, it is necessary to make the equipment sufficiently flexible to enable crops requiring high humidity to be stored at other times of the year. Further details of dehumidification techniques are included in the Technical and Design section.

Atmosphere control equipment

It has been shown by experimentation and commercial practice, that a modified atmosphere is beneficial in prolonging the storage life of some crops. It is important to remember that stored crops respire, using oxygen and producing carbon dioxide. Any attempt to modify the store atmosphere must take this into account. It has been found that different varieties of fruit require different atmosphere regimes if maximum storage life is to be achieved without damage. Three commonly used regimes may be defined as follows:

1. Stores in which the sum of carbon dioxide (CO_2) and oxygen (O_2) concentration equals 21 per cent. For example, 5 per cent CO_2 and 16 per cent O_2 or 10 per cent CO_2 and 11 per cent O_2, the remainder in each case being nitrogen (N_2).

2. Atmospheres where O_2 and CO_2 concentrations add up to less than 21 per cent. For example 5 per cent CO_2 and 3 per cent O_2; the balance 92 per cent being N_2.

3. Atmospheres with less than 21 per cent O_2 and little or no CO_2. For example, 2 per cent O_2 and $0\cdot5$ per cent CO_2: the balance $97\cdot5$ per cent N_2.

The first of these atmosphere types is most easily produced. The CO_2 produced by the produce is allowed to build up to the desired level; it is then maintained at this level by limited ventilation of the store with fresh air. In such a store ventilation is usually controlled by means of a fresh and foul air pipe connected respectively below and above the store circulating fan, the rate of ventilation being controlled by the adjustment of valves at the outer ends of these pipes.

When the storage atmosphere requires both low CO_2 and oxygen levels it is not possible to maintain the conditions by ventilation with outside air. Regimes 2 and 3 fall into this category and so must be provided with scrubbing facilities to remove surplus CO_2. The fresh air ventilation valves are used to control the oxygen level whilst the control of CO_2 is obtained by varying the scrubbing period. Under the third regime where CO_2 levels are kept to a minimum no scrubber control is required; scrubbing is continuous.

CARBON DIOXIDE SCRUBBERS

There are two principle systems by which CO_2 is removed from the store atmosphere. In the first the CO_2 reacts chemically with the absorbent which is discarded when it becomes saturated. In the second the CO_2 forms a lose association with the absorbent which can then be regenerated and used again.

Chemical scrubbing

The most commonly adopted chemical scrubbing method involves the use of dry hydrated lime as the absorbent. The scrubber consists of an insulated cabinet connected to the store by means of 100 mm or 150 mm diameter plastic pipes. The store atmosphere may be circulated through the scrubber either by the action of the store circulating fan or by a separate scrubber fan.

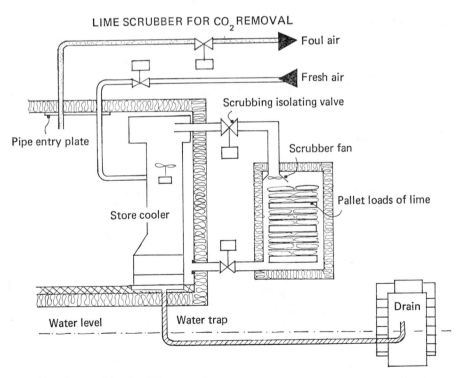

Fig. 14 Lime scrubber for CO_2 removal

Lime quality is important. The lime must have a 90–95 per cent calcium hydroxide content. The highest scrubbing rates will be achieved if individual bags are separated from one another. This can be done by placing timber battens between layers of bags. The scrubber should be fitted with valves so that the scrubber cabinet can be isolated from the store whilst the lime is being changed. In some circumstances these valves may also be useful in controlling the rate of scrubbing.

In some large storage installations it may be considered more convenient to have only one or two scrubber chambers, and to arrange by means of valves for the atmosphere in each chamber to be circulated through the scrubber. In practice it has been found most convenient to connect the chambers with similar atmospheres to one common scrubber. Such an arrangement requires more complex pipe work, but may save on building costs.

Physical scrubbing

In a physical scrubber the atmosphere may be passed over an adsorbing material (e.g. activated charcoal) to which the CO_2 'sticks'. When this is saturated with CO_2 the adsorbing material is removed from scrubbing duty and regenerated.

PHYSICAL SCRUBBING

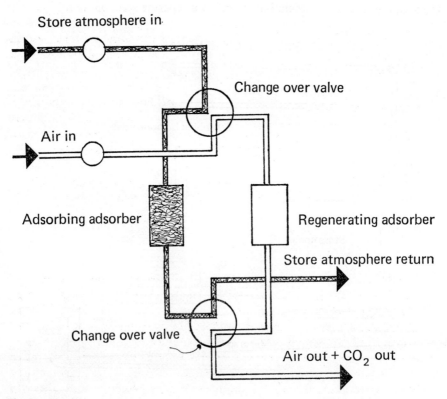

Fig. 15   Physical scrubbing

Physical absorption scrubbing may be considered as a viable alternative to the use of dry lime. Such a system lends itself to automatic control and has the advantage of requiring less labour than the lime scrubber since the active component is regenerated between cycles and does not have to be replaced at frequent intervals. A single scrubbing unit is normally arranged to scrub on a number of stores in sequence. Pipes and control valves are therefore required to enable the scrubber control unit to select the store to be scrubbed.

Typically the equipment consists of two steel vessels containing the adsorbing material. The first of these is connected via control valves to the store to be scrubbed. Store atmosphere is circulated through the scrubber by means of a fan. During this time the other vessel is being ventilated by a separate fan with fresh air from outside to remove any CO_2 that had been trapped by the adsorbing material. After a predetermined period the roles of the two absorbing vessels are reversed, care being taken during the changeover process not to allow O_2 to enter the store.

OXYGEN SCRUBBERS

The normal method of establishing the reduced O_2 concentration in a store atmosphere is to allow the fruit respiration process to use up the oxygen present when the store is closed, and to thereafter control the level by admitting only the required amount of outside air. Satisfactory control can, however, only be achieved if the store is sufficiently airtight. An oxygen scrubber allows the operator to remove any excess O_2 that leaks into the store, thus making the maintenance of conditions easier. A further advantage, particularly in large storage chambers, is that if necessary the storage atmosphere conditions can be quickly recovered after part of the store has been unloaded. Such a system may also be useful for creating the correct atmosphere at the beginning of the season in partially filled stores, though this cannot be generally recommended since the weight loss of fruit in partially filled stores is likely to be excessive. The use of an oxygen scrubber makes possible the rapid production of store atmosphere conditions at the start of storage. It has been suggested that an improvement in storage life and product quality can be achieved by creating storage conditions more quickly than by allowing the fruit to produce them itself.

Catalytic burners

The principal method of scrubbing O_2 from storage atmospheres is to burn propane in the atmosphere thus converting any O_2 present into CO_2 and water. The CO_2 produced by this process is then removed by the CO_2 scrubbing system. As the O_2 level in the store atmosphere is reduced, it becomes more difficult to keep the propane flame alight. In order to enable combustion to occur at O_2 levels of three per cent a catalyst is required to assist the combustion process. Heat is generated during the combustion process, and this must be removed, normally by passing the products of combustion through a cooling water spray.

Atmosphere generators

As an alternative to scrubbing O_2 from the store atmosphere, a suitable atmosphere may be generated by modifying the O_2 and CO_2 concentrations in atmospheric air. The gas produced is then blown into the store displacing the atmosphere that is already there via the vent pipe.

It is important to remember with a 'flow through' or flushing type system that the displacement of the existing store atmosphere proceeds at a diminishing rate. The number of changes of the empty volume of the store required before the desired O_2 level is reached, will depend on the O_2 concentration of the flushing gas, and the efficiency of mixing within the store. Because of the diminishing rate of change of O_2 concentration towards the end of the process, it will be readily appreciated that small adjustments in O_2 level (to compensate for leaky stores) will require considerable volumes of flushing gas. Atmosphere generators have been devised that are capable of producing directly the required storage conditions. They have, however, not been adopted in the UK because of expense. The most common use of the flushing technique is in the initial establishment of conditions when the equipment used is an alternative to the oxygen scrubber.

Open flame generator

The open flame generator produces a flushing gas which is low in O_2 and high in CO_2. This is achieved by burning propane in a controlled flow of air. By careful control of the volume of air supplied, and the shape of the flame, it has been found possible to produce a gas mixture containing 1–2 per cent O_2, the remainder being N_2 and CO_2. The extra CO_2 produced must be removed by the scrubbing system, and as with the oxygen scrubber the heat generated by the combustion process must also be removed. Care must be taken when using this

OPEN FLAME ATMOSPHERE GENERATOR

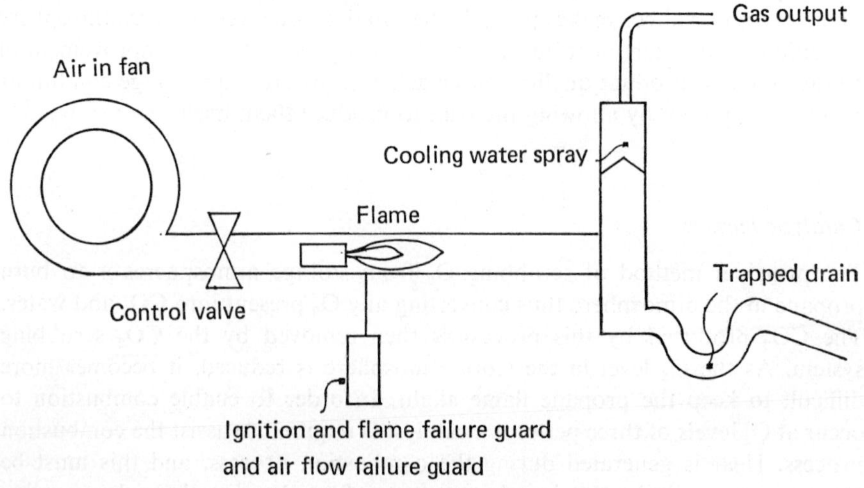

Fig. 16   Open flame atmosphere generator

flushing gas generator that the CO_2 level in the store does not reach a level which may damage the produce. If this happens generation should temporarily be halted, allowing the scrubber time to absorb the excess CO_2. More economic use of this generator may be made by connecting a number of storage chambers in series, the atmosphere from the first being displaced by the generator into the second and so on. Final pull down on each chamber must, however, be made with the generator connected directly to that chamber.

Nitrogen flushing

The injection of nitrogen gas into the store is an effective means of rapidly reducing the O_2 concentration in the store atmosphere. Nitrogen injection can also be used to restore O_2 levels that may have risen as a result of leakage. If, however, the store is very leaky, the volumes of N_2 required to maintain low O_2 conditions may become uneconomic, and may also tend to result in drying of the store atmosphere. The principle advantage of using N_2 to perform the initial reduction of O_2 levels is that very little capital equipment need be installed on the farm and no additional load will be imposed on the carbon dioxide scrubber. Where this technique is employed on large scale storage plants it may become economically desirable to install a liquid nitrogen storage tank.

Control of conditions

In all controlled temperature stores it is essential to have an indication of the temperature at which produce is being held. This serves as a check on the temperature control thermostat, and can give an indication when dangerously low or exceptionally high temperatures exist in the stored produce. Controlled atmosphere stores also require a means of sampling and analysing the storage atmosphere.

TEMPERATURE CONTROLS

The cooling of the store is normally controlled by means of a thermostat. In direct expansion systems the thermostat will stop and start the refrigeration compressor, whilst in indirect systems it will control the flow of coolant to the store cooler. The instrument should be selected to have a differential of $\pm$ $0\cdot5°C$. For direct expansion systems a differential of less than this is not recommended, since it will result in frequent starting and stopping of the compressor: this increases wear on the plant. It also makes it more difficult to maintain the desired store atmosphere conditions because of the pumping effect resulting from the pressure changes occurring within the store as the temperature is alternately raised and lowered. Where it is required to store produce at a temperature only $0\cdot5$ to $1°C$ above its freezing point, an additional thermostat should be provided as a safety measure to stop the refrigeration plant if the air discharge temperature becomes too low.

The set temperature scales on thermostats are often small so that it is better to make final adjustments of the thermostat setting on the basis of temperatures recorded by the temperature monitoring equipment.

As an alternative to the traditional mechanical thermostat, electronic thermostats are now available. The chief advantage of such an instrument is that it provides more flexibility both in terms of operating differential and detector positioning. Electronic thermostats are also less susceptible to the effects of ambient temperature variation than their mechanical counterparts.

Thermostat positioning

The positioning of thermostat detecting elements is crucial. It is in general undesirable to site them in the air discharge from the cooler since this tends to lead to short cycling of the refrigeration equipment. It is important to avoid siting the bulb close to external parts of the structure and equally important to avoid positioning the thermostat detector in a dead area where there is little or no air circulation.

For ceiling mounted coolers, thermostat detectors are best sited close to the air return to the cooler. For floor mounted coolers the best position is just to the side of the cooler discharge opening. Where electronic thermostats are used, the detector may be placed amongst the stored product. In this case it is important to choose a representative part of the stack. In bulk stores where refrigeration is used it is probably best to locate the thermostat detecting element in the return air duct to the cooling equipment. Under these circumstances it is important that the instrument is set to give the correct product temperature. Cold air delivered by the cooler will warm up as it passes through the crop and the return duct work so the actual thermostat setting may have to be about 1°C higher than the required store temperature.

HUMIDITY CONTROL

Many methods have been devised for detecting and measuring relative humidity. Unfortunately, the majority of these methods have technical or economic disadvantages when considered for agricultural applications. For this reason humidity control particularly at high relative humidities is likely to be a theoretical rather than a practical exercise.

Two types of controller are available for humidity control. The first type depends on the change in length which occurs in hygroscopic materials, for example hair or cotton, with changes in humidity. The chief disadvantage of this system is that as 100 per cent humidity is approached the rate of change of length of the hygroscopic element decreases. This tends to result in large switching differentials in the high humidity range. The second type of humidity control system is based on electrical resistance or capacitance measurements of hygroscopic materials. This system is more effective at high humidities.

A permanent shift in the calibration of either type of instrument may be experienced if the detecting element becomes wet or contaminated by dust.

Positioning of humidistats

Experimental measurements have shown that the relative humidity may vary considerably throughout a store. It is therefore difficult to select a representative point at which to make measurements. Ideally the detecting element should be placed in a filtered, gently-moving air stream. It is best to avoid turbulent and fast moving air streams particularly for hydrostats using hair or cotton elements.

AUTOMATIC ATMOSPHERE CONTROL

The introduction of atmosphere regimes employing one per cent oxygen makes control more critical. Automatic control of oxygen level can be achieved by linking the oxygen analyser to valves controlling the store ventilation. Monitoring of oxygen levels would still be required on a daily basis. A check for alcohol vapour in the sampled atmosphere would be an additional safeguard, though this may not be possible in mechanically scrubbed stores where activated charcoal is the scrubbing agent.

Measurement of conditions

TEMPERATURE

The simplest and least expensive temperature measuring system is probably the vapour pressure operated, dial reading thermometer. This instrument would typically be used for short-term vegetable stores where an indication of store temperature is a useful management aid.

For long-term storage an electrical resistance thermometer system can be used to measure temperatures. Such a system may consist of a platinum resistance thermometer, thermistors or alternatively, a thermostat detecting element may be used. In each case the principle of operation is the same. The electrical resistance of the sensing element varies in proportion to store temperature. The reading equipment is calibrated directly in terms of store temperature.

Positioning of thermometers

In short-term stores where the vapour pressure thermometer is used the sensing bulb should be placed next to the thermostat bulb, just out of the direct air discharge from the cooler.

In long-term stores where electrical temperature measuring systems are used the sensing element can be placed in amongst the produce so that they more nearly detect produce temperature and not store temperature. It is usual to provide two sensing elements for the first 50 tonnes of storage and then one additional sensor for each further 100 tonnes. Where there are two elements one is placed near the top of the stack of bins whilst the other is placed near the floor.

Thermistor temperature measuring equipment is normally employed for relatively low value crops since the reading equipment tends to be less expensive than that for platinum resistance thermometers. The number of locations at which the temperatures in a bulk store should be monitored depend on the shape and size of the stack. In general, one temperature detector for every 50–100 tonnes of produce would be considered adequate. Detectors should be located in areas of the store where extremes of high or low temperature are expected. For example the highest temperatures in a bulk stack may be expected within 450 mm of the top surface, whilst lowest temperatures may occur near exposed outside walls.

Calibration of temperature measuring equipment

The calibration of all temperature measuring equipment can be checked by immersing the detecting element in a mixture of crushed ice and water. In such a mixture the reading instrument should register 0°C. As an additional safeguard, it may also be advisable to immerse the detectors in a flask containing water whose temperature has been measured with an accurate, mercury-in-glass thermometer. During such checks it is important to leave the detectors immersed for 10–15 minutes before attempting to read the temperature. If equipment is checked in this way at the beginning of each season one can be reasonably sure that the temperature readings obtained are representative of the conditions that exist in the store. Many temperature measuring instruments contain their own power supplies in the form of a dry battery. In such cases it is important to make a regular check of battery condition since deterioration of the battery will result in incorrect readings.

HUMIDITY

The simplest method of measuring relative humidity is to use a wet and dry bulb thermometer. This instrument consists of two exactly similar thermometers, the bulb of one being kept wet by enclosing it in a muslin wick that is supplied with pure water from a reservoir. The relative humidity is obtained by reading the thermometers and referring to a table.

The main disadvantage with this method is that the development of full wet bulb depression depends on a fairly high level of ventilation, and when storage conditions are close to 0°C there is a danger that the true wet bulb will be below freezing point.

The measurement of relative humidity in the store is difficult, particularly when it is close to saturation and measurement can only be recommended in situations where either adjustments to the cooling plant may result in improved conditions, or where it is suspected that the wrong operating humidity is the cause of problems experienced during storage. In stores designed to operate at high relative humidities it is important to ensure that the correct equipment is installed at the design stage.

ATMOSPHERE ANALYSIS

Where produce is being held in a modified atmosphere, it is necessary to be able to regularly analyse the composition of the atmosphere so that if it varies from that which is desired appropriate control action may be taken. Analysis techniques may be divided into two groups. Firstly those where the composition is determined by chemical means and secondly those which depend on the physical properties of the gas mixture. Generally speaking chemical methods if used carefully provide a higher degree of accuracy but are slower than physical methods. Chemical methods are normally much less expensive than physical ones but require a constant supply of fresh chemicals if accuracy is to be maintained.

Chemical analysis

Chemical analysis of store atmospheres can be performed using an *Orsatt* apparatus. The principle of operation is to draw a sample of known volume into the instrument, to isolate it, and by chemical means absorb the constituent whose proportion it is desired to know. When the absorption reaction is complete, the change in volume at constant pressure represents the proportion of the gas present in the original sample. It is important to note that the chemicals used in the analysis equipment are dangerous and it is probably better to have them made up by a dispensing chemist.

The absorbent used for determining carbon dioxide is 20 per cent potassium hydroxide by weight. A fresh solution of this strength can be made by dissolving 30 g of potassium hydroxide in 50 ml of distilled water. Great care is required since the solution is very caustic, and considerable quantities of heat are evolved as the solid potassium hydroxide dissolves in the water.

The absorbent required for oxygen is alkaline pyrogallol; this reactant can be made up as follows:

75 g of potassium hydroxide are dissolved in 125 ml of distilled water. Again, care must be taken because heat is evolved. In a separate container, 5·5 g of pyrogallol are dissolved in 25 ml of water. As soon as the pyrogallol has dissolved the solution must be added to the 125 ml of potassium hydroxide solution.

The mixture must be put into the absorbing vessel of the instrument or into a ground-glass stoppered bottle as soon as possible. Stale chemicals should never be used for analysis since they are likely to give inaccurate results and the time required for complete absorption of the constituent gas will be extended.

The degree of skill required in operating an *Orsatt* apparatus must not be underestimated. Whilst an experienced operator will obtain results that are probably more reliable than those achieved by using physical analysis, the novice may have difficulty in obtaining repeatable results. Because of the time taken to make an analysis using chemical methods such a system cannot be recommended where there are more than two or three chambers requiring twice daily testing. Nevertheless it is useful to have an *Orsatt* so that in the event of instrument or power failure, continued analysis of the store atmosphere is possible.

Physical analysis

The physical properties upon which the analysing techniques are based are thermal conductivity and infra red absorption for CO_2. Oxygen measuring equipment is based on the gases magnetic properties. When physical methods of analysis are used it is desirable to have a means of standardising the equipment. The simplest method is to provide a bottle of sample gas of known composition. The most convenient mixture is 95 per cent N_2, 5 per cent CO_2. Both oxygen and carbon dioxide analysers must be calibrated at two different points in their range, one of which will normally be zero. The oxygen instrument can be set to zero using the sample gas and its span may be adjusted using ambient air (21 per cent). The carbon dioxide instrument will be set to zero using ambient air and its span will be adjusted using the five per cent CO_2 in the sample gas.

Carbon dioxide analysis

The choice of instrument for CO_2 analysis will depend on the proposed atmosphere composition. A cathorometer of the thermal conductivity type is suitable for the whole range of conditions found in ventilated stores and can be used successfully when the atmosphere contains 5 per cent CO_2, 3 per cent O_2. True readings for the latter case, however, must be obtained from a special conversion table. When the O_2 level falls below the 3 per cent and the CO_2 level below 5 per cent the cathorometer becomes increasingly unreliable.

If accurate analyses of CO_2 levels below 5 per cent when O_2 levels are below 3 per cent are required an infra red analyser should be used.

Oxygen analysis

For oxygen measurement in atmospheres containing 5 per cent CO_2 and 3 per cent O_2 the cathorometer can again be used. The method employed is first to read the CO_2 level, then pass the sample of gas through a carbon furnace where all the O_2 present is converted into CO_2, the resultant gas mixture is then re-analysed and the results converted into true O_2 and CO_2 levels using a conversion table provided with the instrument. When the storage atmosphere contains 2 per cent O_2 and less than 3 per cent CO_2 the cathorometer is not sufficiently accurate in its determination of O_2 for safe storage.

Under these conditions a magnetic oxygen analyser should be used since it will give sufficient accuracy over the whole range of O_2 concentrations likely to be encountered in storage work.

When storage atmospheres containing 1 per cent O_2 are being used, serious consideration must be given to the use of automatic control of the O_2 level.

Management of stores

Successful storage of any produce requires a good technical knowledge of the way in which the store and its ancillary equipment should be operated. Storage installations are composed of many separate parts, often installed by separate contractors; consequently it is unusual to find a set of operating instructions for the user which cover all aspects of the equipment. Staff training is an important part of store management. In view of the high value of the stored produce, a small amount of time and money spent on making sure that the operator knows how the system works will be amply repaid.

STORE LOADING

The rate at which a store will be loaded depends to a large extent on the rate at which the produce is harvested. Controlled atmosphere stores must be completely loaded in the shortest possible time. In practice, one should aim to fill the store in not more than five to six days. During the loading period fruit will continue to ripen, and since the stage of maturity at harvest is of critical importance for successful long term storage delays at loading time may result in poor storage performance.

The loading rate for vegetable stores and fruit stores where no atmosphere control is contemplated should be as rapid as possible but care must be taken not to overload the refrigeration plant since if this happens cooling time may be considerably lengthened. Where a number of cool stores are being filled, it can be advantageous to distribute the daily loading between the stores. Such a scheme will enable the efficient use of all the refrigerating capacity available. Where produce is harvested for long-term storage during the autumn, it is often possible to reduce the refrigeration requirement by standing the boxes in the open overnight and loading them during the following morning. The principal disadvantage of this technique is the increased handling of produce though this may be unimportant where fork lift equipment is used. An alternative system of pre-cooling is to stack each day's loading in an empty store subsequently moving it into its final location the next morning. Some store layouts lend themselves to the construction of a refrigerated passage way onto which individual storage chambers open. Such a layout, in addition to providing extra short-term cool storage capacity, may lend itself to being used as a pre-cooler for the produce that will be loaded into the long-term storage chambers.

Whichever method of cooling is used it is important that the store structure should itself be pre-cooled. It is good policy to run the refrigeration plant for two to three days before loading commences. It is also important to start cooling produce immediately it is loaded. The refrigeration plant should be kept operating throughout the loading period.

Stacking

The use of mechanical handling equipment has led to a different approach to the stacking of produce in store. When stores were hand loaded the most efficient air distribution was achieved by filling the store tight from wall to wall and circulating the air from top to bottom. The introduction of mechanically handled storage units has led to a considerable increase in unit size, typically the bulk bin and the pallet load of small containers. In many cases the construction of bins and containers is such that very little air will pass through the produce as a direct result of the store air circulating fans. Under these circumstances the majority of the heat is removed from the produce as a result of convection occurring within the bin. The function of the store air circulation system in these circumstances is to ensure that the warm air produced by each bin is quickly removed and replaced by a supply of cold air. For this to happen efficiently the air being circulated in the store should pass down a gap between adjacent rows of bins. In practice it is difficult to stack bulk bins so that they are touching, it is normal practice to allow 50–75 mm on each of the plan dimensions of the bin for stacking. This gap will provide an adequate air path for efficient circulation, but it is important that there should not be large variations in the size of the gap. Where stores are expected to cool produce quickly and within a specified time it is of the utmost importance that the accuracy of spacing containers is observed. Failure to do this will result in considerably longer cooling times. Where stores are only partially filled, it is important to ensure that as much of the circulating air as possible passes through the stack of produce. Where a floor mounted cooler is used, stacking the produce around it will help to ensure this. If it is known in advance that the store will not be completely filled, better results would probably be achieved by reducing the loading height in the store. Because of weight loss and atmosphere control considerations, it is undesirable to operate stores in a partially loaded condition for long periods. These problems can to some extent be overcome by the use of plastic film sleeves that cover the sides and top of the cooled pallet. The use of such sleeves has been found to considerably reduce water loss from produce, and provided that the quantity of material covered is not too large (i.e. not greater than a single pallet) and the produce has been effectively cooled throughout, temperature variations within such a sleeve will be small. This technique is particularly useful for short and medium term storage of vegetable crops where it is unlikely that stores will be full.

Where a large store is to be used for rapid cooling of relatively small quantities of produce, for example strawberries, forced air circulation between the trays of fruit can be achieved by the use of portable supplementary fans within the store. An arrangement whereby the trays of fruit are formed into a tunnel with a closed end and the supplementary fan is used to blow air into the tunnel has been found effective. Stores with ceiling mounted coolers may be adapted for high speed cooling by creating a plenum chamber between the produce and the store wall and allowing the cooler circulating fan only to draw air from this space.

STORAGE OF MIXED PRODUCE

On a holding where a number of crops are produced it is inevitable that at some stage there will be a requirement to hold more than one kind of produce in a single store. For short-term storage there are generally no problems provided the following guide lines are observed;

1. The store operating temperature must be set so that produce requiring the highest storage temperature will not be damaged. For example, if it were required to store runner beans and lettuce together, a storage temperature of 7°C will have to be selected to avoid damage to the beans. Such a storage temperature would inevitably lead to reduced storage life for the lettuce, since this would normally be held at 1°C.

2. Products with a powerful aroma should be stored separately; if this is not done there is a danger that less strongly flavoured commodities that are held in the same store may become tainted. In some cases it is possible for the store structure to absorb taint and pass this on to subsequent crops. It should be routine practice to thoroughly ventilate stores that have been used for crops such as onions, to ensure that all traces of the odour are removed. Particular care is needed when stores are used out of season for dairy produce or eggs. Where stores are used for potatoes that have been subjected to chemical sprout suppressants, thorough ventilation of the chamber after storage is essential since some of the chemicals may become absorbed onto the store structure and cause damage to subsequent crops.

3. Ethylene gas is important in the initiation and continuance of the ripening process in stored produce. Vegetables and fruit tend at certain stages of ripeness to produce large quantities of ethylene; the results therefore of storing ripe and unripe produce together will be to accelerate the ripening of the unripe crop. Ethylene also has a very damaging effect on flowers, and causes a serious reduction in their vase life. For this reason it is inadvisable to store any ethylene producing crops with flowers. Certain types of fungal decay result in the production of ethylene; it is therefore important to avoid harbouring such sources in stores.

STORE RECORDS

The importance of keeping a record of temperature and for controlled atmosphere stores, gas concentrations, has been outlined. These records are

STORE RECORD

CHAMBER NUMBER _ _ 1 _ _ _ _   CROP _ _ Cox _ _ _

STORAGE CONDITIONS   TEMP 3·3°C CO_2 5% O_2 3%

DATE		TEMP.		CO_2 %	O_2 %	SUCTION GAUGE	DELIVERY GAUGE	REMARKS
		UPPER °C	LOWER °C					
Dec 1	A.M.	3·5	4·0	5·5	3·6	Plant	off	
	P.M	3·6	3·8	5·5	3·7	27	90	
Dec 2	A.M	3·4	3·6	5·6	3·6			
	P.M.	3·2	3·4	5·6	3·8			
Dec 3	A.M.	3·3	3·5	5·8	3·8			Fan reversed
	P.M.	3·4	3·3	5·7	3·8	25	85	Fresh air reduced
	A.M.							

Fig. 17   Store record

likely to give the first indication of trouble, both with the plant and the stores produce. If the quality of produce at unloading time is unsatisfactory, store records can often give an indication of where the fault lies.

For cool stores it will be sufficient to record the store temperature once a day, whilst for controlled atmosphere stores the temperature and gas concentrations should be read and logged twice a day. In addition to these basic recordings it is helpful to record loading and unloading dates, the dates of lime changes, and any irregularities such as power failures etc. If the refrigeration plant is operating at the time store records are being taken it may be helpful to note the suction and delivery pressures at the compressor. These values will vary throughout the season but if the plant gives trouble they may provide a useful indication to the engineer as to where the trouble lies.

ANNUAL INSPECTION AND PRE-SEASON CHECKS

Before the store is put into use the instruments, plant and structure should all be inspected. The instruments used to monitor conditions in a controlled atmosphere store are the only means by which conditions inside the store during the storage period can be determined. It is therefore very important that they are accurate. The best way of ensuring accuracy is to arrange for contract maintenance of the equipment, including an annual overhaul of the complete instrument panel. A similar contract maintenance scheme for the refrigeration plant is also desirable, particularly if a pre-season check can be arranged so that any faults that have occurred can be put right before the plant will be required to work at maximum load. Controlled atmosphere stores should also be subjected to a gas tightness test. Any leaks that are found can then be repaired before the crop is loaded. Details of a suitable pressure test are given on page 113.

SAFETY

The atmospheres used for prolonging the storage life of produce are incapable of supporting human life. Operators must be made aware that under no circumstances may they enter a controlled atmosphere chamber, unless they are equipped with a self contained breathing apparatus. When stores are opened for unloading it will be necessary to run the store fan with the doors open and the fan hatch removed for at least 30 minutes before unloading is commenced. It is advisable to check the store atmosphere with a portable analyser or a candle before entering the store.

Technical and design section

The following section gives more detailed information on the design of the various systems described in the previous section.

The refrigeration system

COOLING COILS

The selection of the correct cooling coil for any storage application is one of the main factors that will influence the relative humidity of the storage environment produced. Where high humidities are required it is important to size the cooling coil for the smallest practical TD.* (Selections should be based on a TD of between 3 and 5°C.) Such a selection inevitably leads to a coil with a large surface area. Where copper tubes and aluminium fins are used the primary to secondary surface area ratio should be around 1–18. If galvanised steel fins and tubes are used the ratio should be around 1–11.

The number of rows of tubes in the direction of air flow also affects the coils performance. For high humidity applications a deep coil having six or more rows of tubes is generally satisfactory. The face velocity of the air passing through the cooling coil also affects the coils performance. Typical design figures are 2–3 m per second. Such a velocity will give rise to a pressure drop in the air flowing across the coil of 60–80 Pa.

Fin spacing affects the resistance to airflow offered by the cooler. Restricted airflow results in a reduced efficiency of the cooling system. In stores that will run at temperatures above 4°C a fin spacing of 4 mm between fins is acceptable. For store temperatures below 4°C fins should be spaced at 6 mm apart to allow for a certain amount of frosting to occur before the airflow is severely restricted. It is important to avoid obstructions before and after the cooling coil as these may give rise to poor air distribution. The effect of poor distribution is to reduce the effective surface area of the coil resulting in lower humidities than were originally intended in the design.

Some storage applications require low humidities in the store atmosphere rather than high ones. If suitably selected and controlled, the cooling coil can act as a dehumidifier to provide the required conditions. The dehumidification process requires a low coil temperature. Low cooler temperatures will result

*TD is the difference between the temperature of the air entering the cooler and the temperature of the evaporating refrigerant.

from choosing a coil with a small effective surface area. Where a store must be capable of running at both high and low humidities at different times of the year it is possible to valve off part of the evaporator during the period when low humidities are required and to switch to an alternative expansion valve.

The effect of restricting the amount of cooling coil that is operative may result in air temperatures below those required in the store. Under these conditions it may be necessary to reheat the air after it has passed through the cooler. Reheating will ensure the correct storage temperature is maintained and will lower the relative humidity at the same time. An alternative means of controlling relative humidity without using reheat is to provide a by-pass for the cooling coil. In this situation a proportion of the air circulating in the store is allowed to by-pass the cooling coil and is then remixed with the fraction of air that has passed through the coil. If the correct proportion has been by-passed the resulting mixture delivered from the conditioning equipment will have the required temperature and humidity.

Where a store has been designed to operate at high humidities and is required to perform at a lower humidity some improvement in conditions may be obtained by increasing the super heat setting of the expansion valve. This will have the effect of limiting the area of the coil that is actually used.

Where an installation is to be designed specifically for low humidity operation the cooling coil should be selected to have a large TD. (A typical TD would be in the range 10–15°C.)

HUMIDITY/TEMPERATURE CONTROL SYSTEMS

On some occasions (e.g. onion storage and the conditioning and storage of flower bulbs) it is necessary to control both temperature and humidity. Even when these products are dry they tend to give off small quantities of water vapour and so the control of humidity tends to be one of reducing humidities to the desired level whilst still maintaining accurate temperature control. The equipment required for this operation would consist of a refrigerated cooling coil followed by a heater bank and in some circumstances a humidifying device. It is normal to circulate the air first through the cooler, then the heater and finally through the humidifier. In this application the cooler has the dual function of cooler and dehumidifier. Dry bulb temperature is controlled by a three-position thermostat. In the first position, when the room temperature is low the heating would be on. The second position is not used. The third position, when the temperature is high, causes the cooling coil to operate. A three-position humidistat is used to control the humidity. In the first position, when humidity is low the humidifier is operated. The second position is not used. In the third position the cooling coil is brought into operation, in this case as a dehumidifier. A schematic diagram of these controls is shown in Fig. 18.

REFRIGERATION CAPACITY CONTROL

The need to match the refrigeration capacity to the store equipment has been shown to be important if high humidity conditions are to be maintained. In a simple direct expansion system the evaporating temperature depends on the design of the system (cooling coil size, compressor capacity etc.) and on the

BLOCK DIAGRAM OF COMBINED TEMPERATURE
AND HUMIDITY CONTROL

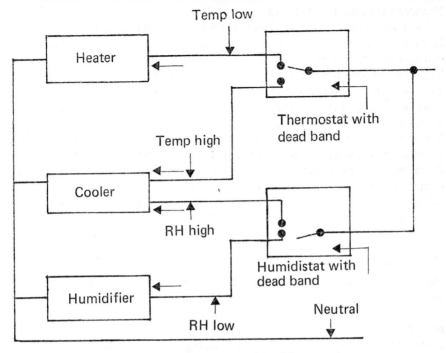

Fig. 18   Block diagram of combined temperature and humidity control

operating conditions (heat flowing to the refrigerating plant and the air
temperature as controlled by the store thermostat).

The simplest and most frequently adopted method of capacity control is
simply to allow the refrigeration plant to be turned on and off at the control
of the thermostat. The conditions existing during the on cycle may lead to an
undesirable degree of desiccation particularly where frequent cycling occurs for
relatively short periods and the evaporator is 'pumped down' to low pressure
by the compressor between cooling cycles. A better match of compressor
capacity and store requirement may be achieved by reducing the capacity of the
compressor. On some machines this may be done by unloading some of the
compressor cylinders. Where cylinder unloading is not possible but the machine
is driven by a separate motor it may be possible to use a two-speed electric motor
thus allowing the compressor to be driven more slowly when unloading is
required. Alternatively the driving pulley size may be altered on machines
where belt drive is used.

The pressure existing in the evaporator controls the temperature at which it
operates. By fixing the evaporator pressure, the effective capacity of the
evaporator will be limited. This occurs because as store temperature approaches
evaporating temperature the effective TD will be reduced. Evaporator pressure
may be controlled by means of a back pressure valve. This valve maintains a
constant pre-set pressure in the evaporator whilst the pressure at the compressor

suction is allowed to fall. Under these circumstances, compressor capacity is reduced by allowing the machine to work against a much lower suction pressure than normal. Even so it is not normally possible to reduce the system capacity by more than about 30 per cent.

The limitations imposed by low suction pressures on capacity reduction can be overcome by providing the compressor with an artificial load. If this is done capacity reduction may continue to 10 or 20 per cent of full load conditions. The artificial load for the compressor is provided by allowing some of the hot gas from the delivery side of the compressor to return to the compressor suction. There is however a penalty for adopting this system since the machine must run for much longer periods to control store temperature. Extra cooling effect that is generated by the compressor is wasted. Since the running costs for storage installations are generally low by comparison with the value of the produce in the store it may be possible to show an economic benefit from the use of such a control system where it can be demonstrated that the weight loss from the stored crop will be less and that an increased proportion of it may be sold in a higher grade.

CONDENSER CONTROL

Other components in the refrigeration system have an indirect effect on the performance of the evaporator. The most important of these is the condenser. Air cooled condensers are designed to operate satisfactorily with the highest expected ambient air temperatures. The temperature control equipment, however, is expected to operate effectively with both high and low outside temperatures. The effect of low outside temperatures on the condenser is to reduce the condensing pressure. In severe cases this may result in a reduction of plant capacity and to some extent a lowering of evaporating temperatures. Many modern air cooled condensing units are provided with a system of condenser stabilisation. The simplest and most frequently employed method is to allow one of the condenser fans to be turned on and off under the control of a pressure switch which detects the compressor discharge pressure. An alternative to the on/off control is to employ a variable speed fan, the speed of which is controlled by sensing the compressor discharge temperature. Another method that may be adopted to control the condensing pressure is to fit thermostatically controlled blinds to the face of the cooling coil. The effect of these blinds is to restrict the airflow through the coil when ambient temperatures are low. The position of the blinds is adjusted by means of a simple thermostatic element which is placed where it will detect ambient air temperature. Condenser stabilisation may also be achieved by restricting the flow of liquid refrigerant from the condenser. A valve placed in the condenser outlet restricts the flow of liquid in response to pressure changes in the condenser. The effect is to maintain a constant condenser pressure over a wide range of operating conditions.

REFRIGERATION GAUGES

Where capacity control equipment is fitted either to the evaporator or the condenser it is important to be able to observe the condensing and evaporating

110

pressures. Gauges fitted to the delivery and suction lines will usually give the operator warning of malfunction in the refrigerating system and enable rapid diagnosis of faults to be made.

The rate at which air is circulated in the store is usually a compromise between the requirements of the cooling equipment and the crop. The rate of water loss from the crop is dependent both on air speed and vapour pressure deficit. At high circulation rates and high air speeds it is possible to maintain a uniformly small vapour pressure deficit. At low circulation rates and low speeds the effect of velocity on moisture loss will be reduced but the vapour pressure deficit throughout the store will tend to be larger than was the case at high speeds.

For a given cooling capacity high circulation rates give small temperature differences between air entering and leaving the cooler; conversely low rates of air circulation give large temperature differences between air on and air off. In addition, the low air circulation rates result in lower evaporating temperatures in a simply controlled direct expansion system. The result of this lower evaporating temperature is to cause relatively more moisture to be condensed from the atmosphere, hence increasing the vapour pressure difference between the cooler and the produce. A further result of low air circulation rates is the possibility of delivering air at dangerously low temperatures. With a simple direct expansion cooling system it is impracticable to vary the air circulation rate whilst the cooler is operating. It is therefore common practice to run the circulating fan at full speed when the cooling plant is operating and to reduce this to half speed when the compressor is off. If back pressure control valves are fitted to the evaporator the tendency of the refrigeration system to low evaporating temperatures under low airflow conditions can largely be prevented. This makes it possible to reduce the air circulation rate for the whole of the storage period.

Where a simple refrigeration system has been installed using a single speed fan, conditions may be improved by arranging for it to run intermittently during compressor off periods. (15 minutes per hour, every hour would probably be adequate.) Where cooling installations have a number of circulating fans designed for single speed operation it may be possible to reduce the number of fans operating during the 'compressor off' periods hence achieving the required reduction in circulation rate. The rate of air circulation over the produce controls the rate at which heat is removed from it. It is therefore important that during the cooling period reasonably high air circulation rates should be maintained. *Typical air circulation rates in fruit stores lie in the range 30–40 times the empty volume of the chamber per hour. Short-term vegetable stores, however, will require higher circulation rates; 60–70 times the empty volume of the chamber is suggested.* The higher circulation rate is necessary since the cooling period will usually be shorter. For long-term storage once cooling has been completed the air circulation rate can be reduced since there is very little heat to be removed from the produce. The acceptable lower limit for air circulation rate is that which gives a sufficiently even temperature distribution throughout the store. It is normal practice to reduce air circulation rate in fruit stores to 15–20 times the empty volume of the store per hour and for vegetable stores it is recommended that the rate be dropped to 20 times the empty volume of the store.

For bulk stored crops where refrigeration is used the air circulation rate can be based on a maximum allowable temperature difference between cooler delivery and return air temperatures. Calculated flow rates would normally be based on maximum load conditions and a temperature difference of 1–2°C. For most bulk stored crops this will result in air circulation rates of 0·009–0·013 m^3/sec tonne.

Ambient air cooling

The effective use of ambient air cooling systems depend on effective automatic control of the ventilating fan. Essentially the system consists of a high capacity ventilating fan and an air distribution system. When conditions are favourable for cooling the building is ventilated with outside air. When the produce is sufficiently cool or conditions are unfavourable for cooling the store air is recirculated at intervals to prevent temperature gradients developing. A block diagram of the control system is shown in Fig. 19. The differential thermostat compares store temperature with ambient air temperature and allows the fan to be switched on when ambient air temperature is lower than store temperature. The store thermostat allows the fan to be switched on only when the produce requires cooling. The frost thermostat acts as a safeguard to prevent freezing air being delivered by the fan.

In order that the ventilation system can be completely automatic it is necessary to be able to automatically select either fresh air or recirculated air. This function is provided by means of the proportional controller and either a pair of louvres or a motorised flap valve.

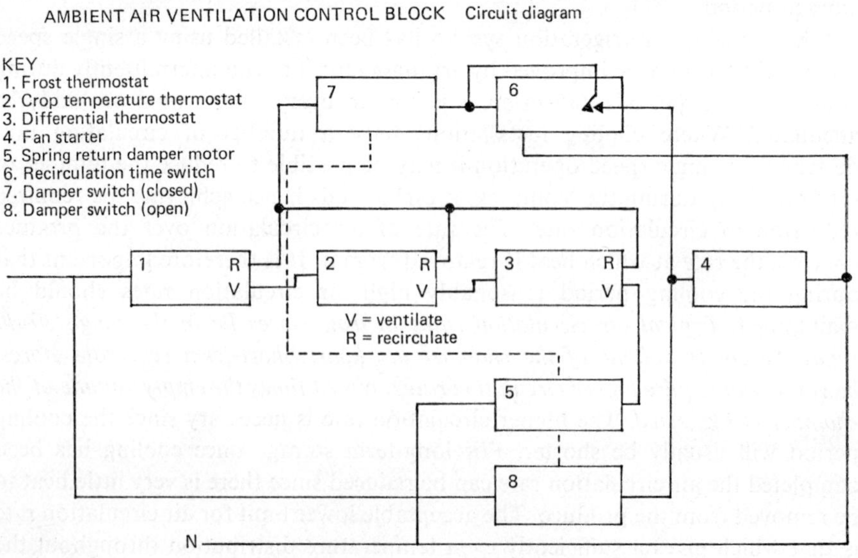

AMBIENT AIR VENTILATION CONTROL BLOCK   Circuit diagram

KEY
1. Frost thermostat
2. Crop temperature thermostat
3. Differential thermostat
4. Fan starter
5. Spring return damper motor
6. Recirculation time switch
7. Damper switch (closed)
8. Damper switch (open)

V = ventilate
R = recirculate

Fig. 19 Ambient air ventilation control block circuit diagram

112

Where this system of temperature control is adopted for crops such as red beet and carrots there is a danger that the ventilating air may be too dry, hence causing serious desiccation of the crop. Where buildings are to be used for the storage of these crops it is strongly recommended that a humidifying water spray is used whenever the fan is operating. When a humidifying system is used some kind of droplet eliminator should also be installed to prevent the crop being wetted. It is also important to ensure that any temperature sensing elements of the control system do not become wetted since this could lead to inaccurate control.

Testing of controlled atmosphere stores

If a store is to be used for controlled atmosphere storage it should be tested to determine whether the structure is sufficiently gas-tight to enable the required conditions to be maintained. The test procedure consists of increasing (or decreasing) the pressure within the empty chamber and determining how long it takes for equalisation of pressure between outside and inside to occur. The leakage rate for a store depends on the total area of holes in the gas lining and the pressure difference existing between the inside and the outside of the store. Wind pressure and barometric changes both result in pressure differences which will cause the store to leak. In addition to these pressure changes, the operation of the refrigeration equipment and some carbon dioxide scrubbers will also cause leakage to occur.

When considering a low O_2 atmosphere the maximum acceptable rate of leakage is that which corresponds with the minimum uptake of O_2 by the produce. If O_2 leaks into the store at a faster rate than this, the O_2 level will tend to rise and it will be impossible to maintain the correct conditions.

Satisfactory results for low O_2 storage have always been achieved in stores where the time taken for the pressure to fall from 187 to 125 Pa has been seven minutes or more. It is recommended that all new controlled atmosphere stores be tested to this standard or better. The following table shows how the test time may be adjusted for maintenance tests when it is proposed to use the store for different atmosphere conditions.

Table 8

Atmosphere analysis

O_2 (%)	CO_2 (%)	Test Time
2·5	0·5–1·0	7 min
3·0	5·0	6 min 50 sec
11·0	10·0	2 min 22 sec
16·0	5·0	1 min 10 sec

TEST PROCEDURE

The equipment required to make a store test is a blower (or vacuum cleaner) to raise (or lower) the pressure in the chamber, and a manometer to measure the pressure in the chamber. Before testing make a careful inspection of the store. Look for cracks in corners, between walls and floor, ceiling and walls round fan hatches, pipe entry plates and door frames. Examine the wall closely with the aid of a strong lamp to see if there is any cracking or crazing of the gas seal. Often common sources of leakage are cable entries and even the cables themselves, duct fixing points, and the store lining behind duct work and coolers. Any defects that are found at this stage should be made good before the pressure test is applied.

The store should be closed up as it would be when full of produce. Points to bear in mind at this stage are:

Check that the drain is full of water,

Make sure that the scrubber (if one is fitted) is correctly sealed and open to the store.

The blower or vacuum cleaner is connected to the store vent pipe. The manometer is connected to the second vent pipe by means of a bung and a small diameter tube; all other vents to the store being closed.

Start the blower and watch the manometer. When the pressure in the store has risen to about 200 Pa stop the blower and shut the ventilation valve. Never apply more than 250 Pa of pressure as this may damage the store. Positive

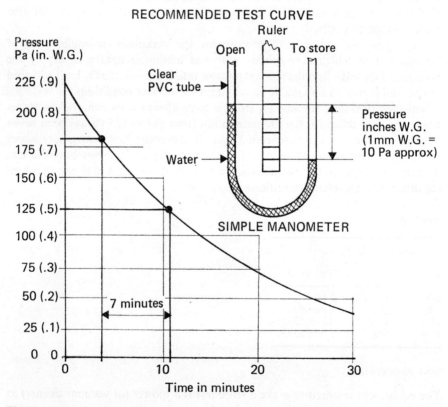

Fig. 20   Recommended test curve

114

pressure tests are preferred on older stores where there is a danger that the gas lining may 'be sucked off' the wall if a vacuum is applied. New stores should give equally good results whether tested under vacuum or pressure. Any differences that are found may be attributed to the improved sealing of doors and hatches when a vacuum test is applied.

Note the time taken for the pressure to fall from 187 to 125 Pa. This time should not be less than seven minutes for a new store. Fig. 20 shows the decay curve for pressure in a store which conforms to the recommended standard. If the curve is plotted for the store under test, and it lies on or above the standard curve the store can be considered satisfactory. If the plotted curve lies below the standard curve the store must be examined to find the source of leakage. The same test curve applies whatever the volume of the chamber under test. During the test period the cooler and circulating fan should be switched off. It is also desirable to measure store temperature at the start and finish of the test to be sure that no change has occurred. A change of $0 \cdot 5°C$ in a closed store will result in a pressure change of 200 Pa.

LEAK DETECTION

If when the blower is started it is found impossible to raise the pressure in the store and the structure has been carefully examined and found in good condition, look for obvious leaks like open vent valves etc. If there are no obvious mistakes it can be helpful to attempt to pressurise the store whilst at the same time introducing a small amount of refrigerant (e.g. R12) into the suction side of the blower. The escape of air from the store containing R12 can be identified by the use of a refrigerant 'leak detector'. The sort of places to look with the leak detector are around door frames, gas panels and pipe entries.

If the store holds some pressure but cannot be raised to a high enough pressure or leaks too quickly the hole is likely to be small. The method outlined above using R12 is not likely to be so effective because the point of leakage outside the store may not correspond to the leakage point inside the chamber.

An external scrubber can be tested for leaks by making two tests of the store. The first with the scrubber pipes blocked, the second with them open. It can be assumed that any difference in the leakage rate is due to the scrubber.

A number of other methods have been devised for the location of holes in stores; all of them involve someone going inside the chamber while it is pressurised or evacuated.

It is sometimes possible to hear air leaking through small holes. Aids such as stethoscopes and directional microphones can be used to make the source of the sound easier to trace. A simple and effective method is to trace all potential leakage areas with a lighted candle. The air jet produced by a small leak when the chamber is under a partial vacuum is sufficient to deflect the candle flame.

When the test shows about three minutes for the recommended pressure drop all major leaks will have been found. The remaining holes may be too small to affect the candle flame. Under these conditions a solution of soapy water splashed onto potential leakage areas may be used to locate the remaining holes. It is best to use this technique in a systematic way so that the whole store lining is checked.

115

Once the leakage points have been found they must be blocked. Where bituminous linings are used the major holes can be covered by applying bituminous plaster or mastic. Care must be taken that all grease is removed before the repair is started. Household detergent and hot water are most effective for this. Where the store is insulated with expanded polystyrene it is important to avoid solvents that may attack this. Where the original gas membrane has cracked it may help if some glassfibre scrim is incorporated in the repair so that it reinforces the bridge across the crack. Small holes and crazing of the surface can best be treated by several coats of a bituminous based aluminium paint. Alternatively a glassfibre, reinforced plastic membrane may be applied throughout the store. Leaks in stores with metal gas linings can normally be cured by the addition of extra grease. Alternatively a self adhesive, metal foil strip can be stuck over any leaky joints: again care is needed to ensure that all grease is removed before applying the strip.

Design calculations for atmosphere generation

SCRUBBING CAPACITY

The selection of scrubber capacity for a controlled atmosphere installation depends on the rate of CO_2 production of the stored produce under controlled atmosphere conditions and the characteristics of the scrubber. In general as the CO_2 level rises the rate at which the scrubbing system can remove CO_2 also rises. This situation gives rise to a balance situation where the rate of CO_2 production is matched at some CO_2 concentration by the rate of CO_2 absorption. Scrubbing capacity must be selected such that the CO_2 concentration at the balance point is equal to or less than the concentration required in the storage atmosphere. The situation is further complicated for lime scrubbers by the fact that the rate of CO_2 removal by the lime decreases as more of the lime is used up. The result of this is that the CO_2 level at which the crop's production and the scrubber's absorption balance, tends to rise as the lime becomes ineffective. This state of affairs does not occur with physical scrubbing systems involving frequent regeneration of the scrubbing medium.

Scrubber capacity for lime scrubbed stores

The recommended scrubber capacity for low O_2 conditions is that it should hold one 25 kg paper sack of lime for every 1250 kg of store capacity. It is important to remember when calculating the space required for the lime bags that they should be separated from one another so that the maximum surface area is exposed to the circulating atmosphere. In order to ensure good circulation through the scrubber a scrubber circulating fan is recommended. This should have a capacity of 0·006 cubic metres per second per tonne of store capacity.

It is desirable to return the atmosphere from the scrubber to the inlet of the store cooling system. Such an arrangement will normally prevent any marked shrivelling of the fruit which might otherwise have occurred if the return pipe delivers directly into the store.

If the storage atmosphere required is to contain less than 0·5 per cent CO_2 a practical alternative to fitting a lime scrubber is to load the lime directly into the storage chamber. If this procedure is adopted, the quantity of lime that must be placed in the store to ensure CO_2 control for the complete season is one 25 kg sack for every 500 kg of remaining store capacity.

A centralised scrubber may be used to serve a number of chambers, the atmosphere from each being circulated through it. In practice it has been found more convenient to operate all the chambers scrubbed by one common scrubber with the same atmosphere and to connect all the chambers together. Such an arrangement requires more complex pipe work, but may save on building costs.

The scrubbing capacity required in special installations can be determined using the graph (Fig. 21) and the rates of CO_2 production given on page 138, Appendix 4c.

The number of 25 kg bags required is found by determining:

1. the rate of CO_2 production (g/h) by the whole store,

2. deciding the percentage exhaustion of the lime when it will be changed, and

3. the maximum permissible CO_2 level in the store.

From the graph the absorption rate per 25 kg bag (g/h) can be read off for the chosen CO_2 level and percentage exhaustion. This rate is divided into the CO_2 production rate for the store to give the number of lime bags required.

The time between lime changes can be estimated from:

1. the percentage exhaustion of the lime,

2. the CO_2 production of the produce, and

3. the quantity of lime in use (kg).

One kg of lime can absorb 0·59 kg of CO_2 during its conversion to the carbonate. The life of the lime is given by the following equation:

$$D = 24 \cdot 58 LE/P$$

Where D = Time to exhaust the lime (days)
   L = Lime capacity of the scrubber (kg)
   E = Percentage exhaustion at end of scrubbing (decimal)
   P = CO_2 production from the store (g/h)

In general, lime is more efficiently used in a large scrubber than in a small one.

Example. An 80 tonne fruit store is to be run at an atmosphere of 2·5 per cent O_2, 0·5 per cent CO_2. The upper limit for CO_2 is set at 1·5 per cent. The lime scrubber capacity is required for 50 per cent exhaustion of the lime.

The CO_2 production by the fruit is 6 g/tonne h. Therefore production rate from the store is $6 \times 80 = 480$ g/h. From the graph the absorption rate is 9 g/h bag at 1·5 per cent CO_2 for 50 per cent exhaustion. The number of 25 kg bags is given by $\dfrac{480}{9} = 54$.

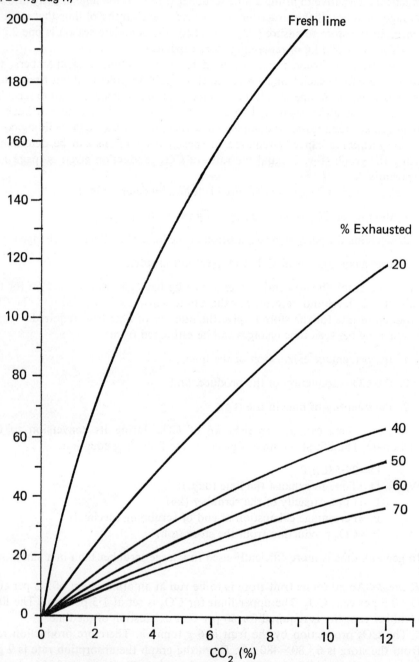

Fig. 21 CO_2 absorption rate per 25 kg bag fresh lime at different CO_2 levels in store

118

The time between lime changes, assuming constant CO_2 production from the fruit, is:

$$D = 54 \times 25 \times 0.5 \times \frac{24 \cdot 58}{480} = 34 \text{ days (5 weeks approximately)}$$

FLUSHING CA STORES

In some circumstances it may be considered desirable to adjust the composition of a CA store by injecting N_2 or CO_2 or a mixture of these gases. When gas is injected into a store some of the original store atmosphere will be displaced. The situation is complicated by the fact that during the injection process the injected gas becomes mixed with the store atmosphere and as a result some of the injected gas is also lost from the store. If it is assumed that perfect mixing of the store atmosphere and the injected gas occurs, and that the void fraction of the store is known, it is possible to calculate the quantity of gas that must be injected to produce the required conditions. In practice there is imperfect mixing of the injected gas and the void space in the store is not accurately known.

An approximate indication of the quantity of nitrogen flushing gas required can be obtained from the following equation:

$$V = A \log_e (C_o/C)$$

Where V = Volume of gas to be injected (m³)
C_o = Initial concentration in the store atmosphere (per cent)
C = Desired concentration in the store atmosphere (per cent)
A = Store void space (m³)

For example, if N_2 is to be used to reduce the O_2 concentration in a 50 tonne apple chamber from 21 to 3 per cent, the minimum quantity of N_2 required—assuming a void fraction for the store of 0.65 and a store volume of 172 m³—would be 217.5 m³ of N_2.

If, as is the case with an open flame generator, the flushing gas contains some O_2, the quantity of flushing gas required to reach a given O_2 level will be greater.

An approximate estimate of the volume of flushing gas required can be obtained from the following equation:

$$V = A \log_e [(C_o - C_a)/(C - C_a)]$$

Where V = Volume of flushing gas (m³)
A = Chamber void space (m³)
C = Final oxygen concentration (per cent)
C_a = Oxygen concentration in the flushing gas (per cent)
C_o = Initial concentration of oxygen in the chamber (per cent)

For example, consider the same 50 tonne store as in the previous example and assume a 2 per cent (0.02) O_2 concentration in the flushing gas (C_a). The volume of flushing gas required would be 329 m³.

The injection of CO_2 into a store is theoretically similar to N_2 injection. In practice, however, it can be considered as a simple case of displacement of an equivalent volume of store atmosphere up to CO_2 concentrations of 20 per cent. Above this concentration it is necessary to treat the problem in the same way as for the first example.

119

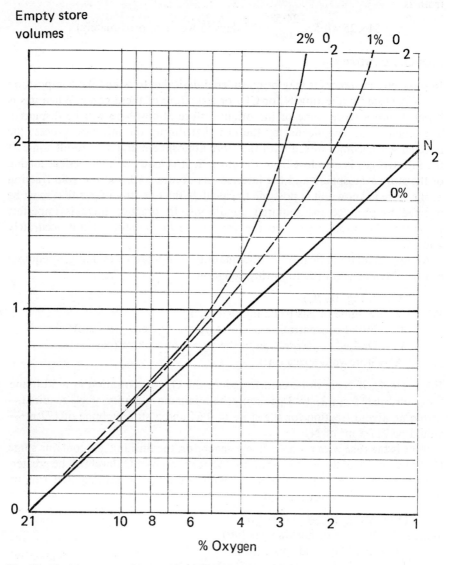

Fig. 22   Flushing gas requirements for CA stores

Refrigeration plant design

The refrigeration requirements for a cooling and storage installation can only be calculated when details of the operating requirements are available. The refrigeration engineer is generally required to specify a maximum plant capacity that will ensure that the required temperature and humidity is maintained within the store at all times when it is in use.

Generally the maximum refrigeration requirement will occur during the time when field heat is being extracted from freshly loaded produce, but there are situations where the maximum load can occur at the end of the storage period. This situation is most likely where long cooling periods are permissible and the temperature at the start of the cooling operation is low. Examples of such situations are long-term onion and potato stores.

It is important not to over-estimate the refrigeration requirements since, apart from the cost of installing a more powerful system, difficulties may be encountered when peak cooling requirement is passed and the cooling plant is expected to operate at one quarter to one fifth of its rated capacity.

CALCULATION OF REFRIGERATED LOAD

The maximum refrigeration load will be made up of the following components:

Removal of field heat.
Heat evolved by product respiration.
Heat leakage through the building structure.
Heat leakage due to air exchange with outside.
Electric fans, brine pumps, electric motors used inside the store and store lights.
A defrost allowance (where it is expected that ice will form on the cooling coils).
A margin to allow the plant deterioration and exceptional conditions.

Removal of field heat

The refrigeration capacity required to remove field heat depends on the loading rate and the allowable cooling time. It will also be affected by loading temperatures and the minimum store temperature.

Under some conditions the rate at which the produce can give up its heat will control the time taken for it to reach storage temperature. Such a situation commonly arises where vegetables are packed into pallet-sized containers and allowed to stand freely in the store. Experimental work has shown that this situation can be improved by ensuring that the cooling air is made to pass through the container rather than round it. If a system of positive ventilation is used, the full benefit of rapid cooling will not be realised unless a very large refrigeration capacity is available.

A further situation in which product cooling rate can have an important influence is in pear storage where the final difference between store temperature and product temperature may only be 0·5°C. In general, it can be assumed that the time taken to cool produce is, with the exceptions outlined, dependent on the refrigeration capacity available. For most conditions, therefore, the cooling requirement is given by:

$$Q = (M\ Cp)\ (TI–TF)/3\ 600\ \theta$$
Where Q $=$ Cooling load kW
M $=$ Mass of produce kg
Cp—Specific heat of produce kJ/kg°C
TI $=$ Initial (loading) temperature °C
TF$=$Final (storage) temperature °C
θ $=$ Cooling time in hours (see Appendices 4f and 4g)

The weight of the load (M) should be divided into produce and container and separate specific heats (Cp) taken for each part.

Loading rates

The values of M and θ that should be used in the calculation of cooling requirement depend on the recommended maximum cooling time for the product and the rate at which the store is loaded.

1. When the recommended cooling time is more than 24 hours, take θ as the recommended cooling time or the expected minimum loading period, whichever is longer. In this case take M as the maximum store capacity.

2. When produce must be cooled in less than 24 hours but more than 12 hours, (assuming the container and packing arrangements will permit such a cooling time) take θ as the recommended cooling time and M as the maximum quantity of produce that will be loaded in a 24-hour period.

3. When produce must be cooled in less than 12 hours, airflow rates and refrigerating capacity must be determined using the results of cooling experiments with the type of package to be used.

Appendix 4g shows the cooling requirements for soft fruit: the table is based on cooling soft fruit where air is directed over the fruit. A minimum airflow rate of 1·57 m³/s 1000 kg, a minimum store temperature of 2°C, and a starting temperature of 27°C are assumed. The times shown are the maximum cooling time (hours) to reach the temperature shown.

For lower starting temperatures, subtract the time shown under the required starting temperature from the cooling time shown for the finishing temperature. The line dividing the table shows the product temperature at which produce heat transfer becomes limiting.

The table can also be used to obtain an indication of the quantity of soft fruit that can be cooled in an existing store. For example: an existing 50 tonne top fruit store is known to have a refrigerating capacity of 10 kW at 2°C, 60 per cent of this will be available for removing field heat. From the table it can be seen that to cool from 26°C to 4°C in three hours requires 11 kW/1000 kg of fruit. The quantity of fruit that may be loaded in three hours is cooling capacity available divided by the cooling required per 1000 kg, $\frac{6}{11} \times 1000$ kg=545 kg.

The average hourly loading rate for the store will be $\frac{545}{3}$=181 kg/hr.

Heat evolved by the produce

All living produce evolves heat. This heat production must be taken into account when calculating refrigeration capacities. The rate at which heat is evolved by produce is a function of the temperature of the produce, its type, its stage of maturity and the degree of damage which it has sustained. Respiration rates also vary between varieties and can be affected by the composition of the store atmosphere (see Appendix 4b).

122

Refrigeration requirements are normally based on average maximum respiration rates at the mean of loading and storage temperatures.

When the cooling requirement is based on a cooling time of less than 24 hours, the respiration load should be calculated in two parts. Firstly, total store capacity less one day's loading at a respiration rate equivalent to storage temperature. Secondly, one day's loading at the mean of storage and loading temperature.

Heat leakage through the structure

The refrigeration load due to heat leaking through the store structure depends on the store temperature, the temperature of the outside surface of the store and the thickness and type of insulation used.

Recommended U values for particular storage applications can be obtained from Appendix 2. The temperature difference across the insulation can, in the majority of cases, be taken as the average monthly maximum less the storage temperature. In cases where the maximum refrigeration requirement occurs at the end of the storage period and is mainly due to high ambient temperatures, it may be desirable to calculate the average heat gain by the structure over a 24-hour period, taking into account the effects of solar radiation. A procedure for carrying out this calculation is described in the Institute of Heating and Ventilating Engineers Guide, Book A, 1970 (from IHVE, 49 Cadogan Square, London, SW1). Since the calculation procedure can be tedious, it is suggested that only the roof is worthwhile considering in this way.

Heat leakage through floors

The rate of heat leakage into the store from the floor depends on the store temperature, the time of year and to a large extent on the length of time that the store temperature has been maintained. The reason for this is that in the initial stages the floor slab and the soil beneath it is being cooled. Once it has been cooled the heat flow rate decreases to an almost constant rate. The rate at which the heat flow from the floor slab falls off depends on the level of insulation.

If it is assumed that the store will be pre-cooled for five days before loading starts and the floor is insulated with 50 mm of extruded expanded polystyrene, a maximum heat flow rate of 9 W/m² can be assumed. This heat flow rate will decrease to less than 6 W/m² after about 20 days. After five days pre-cooling an uninsulated floor may still be giving up heat at the rate of 38 W/m². At ten days this may have fallen to 19 W/m² and may reach 6 W/m² after around 45 days.

The heat flow rate at the edge of the slab is considerably higher than in the centre once steady conditions have been reached. For this reason edge insulation of floor slabs may be considered as an alternative to insulating the whole floor. The majority of the heat leakage at the edge of the floor occurs in a strip about one metre wide. A strip of insulation 600 mm wide × 50 mm thick will reduce the edge effect to an insignificant level. A similar effect may be obtained by extending the wall insulation vertically down into the ground by a similar amount. Where edge insulation is used it should be noted that the initial heat

I

flow rate from the floor will be similar to that for an uninsulated floor. This value must be taken into account when determining the maximum refrigeration load.

Heat leakage due to air changes

Whenever the doors of a store are opened there will be an exchange of store air with outside air. Where doors are opened frequently these air changes may represent a considerable refrigeration requirement.

An approximate indication of the refrigeration requirement to counter air changes can be obtained by assuming a specific heat for air of 1 kJ/kg°C. In most situations this calculation is likely to underestimate the refrigeration requirement since the air entering the store contains moisture, some of which will be condensed by the refrigeration plant, representing an additional load. Where an accurate figure is required a psychrometric table or chart should be used.

Heat gain due to electrical equipment within the store

All the electrical power supplied to equipment that is mounted inside the cooled space will be converted to heat and will have to be removed by the refrigeration equipment.

The most important source of electrical heat within the store will be the air circulating fan. The power input by the fan is given by:

$$Q = \frac{p(\text{total})v}{(e\ 10)}$$

Q = kW
v = volume (m³/s)
p = fan pressure (newtons/m²)
e = efficiency (per cent)

Air circulation rate will in most cases be controlled by the range of temperatures that is acceptable within the store under steady conditions. The minimum temperature variation throughout the store will be limited by the need to avoid excessive fan capacities which result in high running costs and desiccation problems.

Because of the moisture removal by the refrigeration equipment it is safer to quote design limits in terms of enthalpy rather than temperature. The range that will generally be found acceptable is between 2·5 kJ/kg and 5 kJ/kg. Fan capacity may be estimated by summing all the refrigeration requirements (kW) and dividing by the appropriate design enthalpy difference (kJ/kg). The resulting flow rate (kg/s) is converted to m³/s by multiplying by the specific volume (m³/kg). For example a store has a refrigeration requirement of 9 kW (excluding fan heat) and requires a design enthalpy difference of 2·5 kJ/kg with air of specific volume 0·8 m³/kg. The fan capacity is $\frac{9 \times 0\cdot8}{2\cdot5} = 2\cdot88$ m³/s.

In some circumstances it may be justifiable to use a higher fan capacity than would be indicated by the above limits but this would only be the case where short cooling times are required.

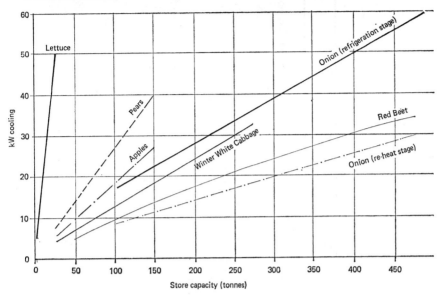

Fig. 23   Typical refrigeration requirements including defrost

Defrost allowance

Refrigeration equipment is designed to operate for 24 hours a day if necessary, but in a direct expansion system it will not be possible to operate the plant for 24 hours per day, because some time will be required to melt ice that forms on the cooler. The following general recommendations for plant running time may be used in assessing the maximum refrigeration load.

Table 9   Defrost allowance

	Store temp.	Evaporating temp.	Plant running time in 24 h.
No defrost	above 2°C	above −1°C	22
Off cycle defrost	above 2°C	below −1°C	16
Power defrost	below 2°C	below −1°C	20

Margin for deterioration

In some circumstances it may be considered desirable to add a safety margin to the calculated load. Such a margin may, to some extent, allow for the inevitable effects of wear in the compressor and other working parts. It may also compensate for uncertainty in the calculation of components of the refrigeration requirement. In no circumstances should the margin exceed ten per cent, and in cases where there is reasonable certainty that the refrigeration requirement has been calculated accurately, no additional margin need be added.

I*

Appendices

Appendix 1

FAULT FINDING CHART FOR EQUIPMENT

Store Conditions	Cause	Cure
Humidity too low (shrivelled produce)	Wrong design of cooler	Larger cooler needed
	Use of dry packing containers	Use non-absorbent packings—soak wooden boxes before use
	Excess air circulation	Use 2-speed fan—reduce fan running time
	Too long storage	Avoid over storage
	Defective insulation	Dry insulation and provide efficient vapour barrier
	No compressor unloading	Unload compressor after cooling is complete
	Wrong super heat setting	Decrease super heat setting
	Too much defrosting time	Reduce number of defrosts given
Humidity too high (bulb and onion stores)	Air leaks (summer)	Make store more airtight
	Evaporating temperature too high	Adjust super heat—use smaller cooler—by-pass some air round cooler
	No re-heat provided	Provide re-heat
	Faulty humidistat	Re-calibrate humidistat
Store temperature too high (fruit too ripe)	Defective thermostat	Replace
	Faulty temperature measurement	Check calibration—check battery if used
	Insufficient cooling capacity	Larger plant needed—reduce loading rate
	Loading rate too high	Reduce loading rate—pre-cool
	Blocked condenser	Clean
	Iced-up cooler	Defrost—increase defrost frequency
	Restricted store air circulation	Re-arrange stacking
	Fan running backwards	Reverse fan
	Belt slip on compressor drive	Tighten or replace belt
	Insulation poor	Check vapour barriers—increase insulation thickness

Store temperature too low	Faulty thermostat Faulty temperature measurement	Replace Check calibration—check battery if used—check location of detectors
Low temperature breakdown (Freezing damage)	Low condensing pressure Vapour pressure thermostat body cooler than bulb in store Leaky refrigerant stop valve	Consult refrigeration engineer—restrict condenser air supply Warm the thermostat with a 15 W bulb Replace defective parts
Poor temperature distribution (Variation in shrivel) (Variation in ripeness)	Poor stacking of produce Part filled store Inefficient distribution of cool air More heat leakage through some parts of the store Not enough air circulating	Stack for even air flow Stack for even air flow Fit ducts—re-position cooler Apply extra insulation—shade the store Check fan is blowing the right way—check stacking—defrost
CO$_2$ level too low (ventilated store)	Too much ventilation Store leaks too much Insufficient produce Windy conditions Leak in sample line Faulty instruments	Close vent pipes Check sealing of doors and hatches—check gas barrier next season Store in air Close vent pipes if wind is expected Sample direct from the store—check joints See 'instrument error'
CO$_2$ level too high (ventilated store) (Brown heart)	Not enough ventilation Produce not cool Seasonal variation in CO$_2$ production Faulty instruments	Open vent pipes—open hatches a crack See 'temperature too high'—open hatch See 'instrument error'
CO$_2$ level too low (scrubbed store)	Too much scrubbing Leaky store (with high O$_2$ level) Leaky sample line Faulty instruments	Reduce scrubbing time—partly close scrubber valves Re-seal store Sample direct from store See 'instrument error'
CO$_2$ level too high (scrubbed store) (Core flush)	Not enough scrubbing time Poor circulation through scrubber Lime exhausted Lime stacked too tight Lime in plastic lined sacks Poor sample of lime Faulty instrument	Increase scrubbing time Fit a circulating fan—re-arrange scrubber pipes Replace lime Put spacers between bags Cut sacks open Request public analyst to check the calcium hydroxide content See 'instrument error'

Appendix 1—*continued*

Store Conditions—*continued*	Cause	Cure
O_2 level too low (scrubbed store) (Alcohol smell/taste to fruit)	Insufficient ventilation	Open vent valves
O_2 level too high (scrubbed store)	Too much ventilation High winds Leaky sample line Leaky store Frequent plant cycling Faulty instruments	Close vent valves Close vent valves when high wind is forecast Sample direct from store Re-seal hatches—use O_2 scrubbing or N_2 flushing to control leaks Increase thermostat differential See 'instrument error'
Instruments		
Apparent instrument error	Incorrect standardisation Unrepresentative gas sample Blocked or leaking sample line Variations in temperature Battery voltage low (thermometers)	Check operating instructions—check standardising gas sample Pump for long enough to clear sample line Clear or repair—pressure test for leaks Keep instruments at an even temperature Replace battery
Plant		
Condensing pressure high	Obstructed condenser coil Dirty condenser coil Fan belt slack High ambient temperature Air in refrigeration system	Remove obstructions Clean Renew or tighten belt Improve supply of cool fresh air Call refrigeration engineer
Condensing pressure low	Low ambient air temperature Lack of refrigerant	Fit condenser capacity controls Check for leaks—charge system—call refrigeration engineer

Symptom	Cause	Remedy
Evaporating pressure low	Iced up cooling coil	Increase defrost time
	Cooling plant too large for load	Unload compressor if possible
	Restricted airflow over cooler	Check fan is working—check coil is not obstructed
	Too high super heat setting	Call refrigeration engineer
	Lack of refrigerant	Check for leaks—charge system
	Blockage in refrigerant line	
Evaporating pressure high	Refrigeration plant overloaded	Reduce loading rate—increase plant capacity
Bubbles in sight glass	Lack of refrigerant	Check for leaks—charge system
Frosting of suction pipe and cylinder heat—clatter from compressor	Liquid refrigerant carry over from evaporator	Call refrigeration engineer—adjust TEV setting (if excessive compressor noise stop the plant—damage may occur)
Excessive frosting of cooler	Too much water in store	Drain dipped produce before loading
	Excessive use of humidifiers	Check efficiency of defrost—increase defrost time
	Insufficient defrost	Check efficiency of defrost—increase defrost time
	Evaporating pressure too low	

Appendix 2a

Thermal conductivity (k)

The 'k' value of a *material* is a measure of that material's ability to transmit heat. It is expressed as heat flow in watts per square metre of surface area for a temperature difference of 1°C per metre thickness (W/m °C).

Thermal transmittance (U)

The U value is a property of an element of a *structure* and is a measure of its ability to transmit heat. It is defined as the quantity of heat which will flow through unit area in unit time, per unit difference of temperature between inside and outside environment (W/m² °C).

Appendix 2b

Heat transfer calculations can be complex and must take account of the degree of exposure of the surfaces, the physical nature of the surfaces and the thermal conductivity value of each component of the structure. Simplified methods are sometimes used taking into account only the values derived from the insulating materials themselves. Approximations of this sort are often made when insulating materials are to be applied on the inside of structural elements. Each case has to be considered separately and the designer must make the appropriate decision in order to determine the thicknesses of insulation required in relation to the equipment and duty. Some k values of common materials are given in this appendix which also lists the U values that should be aimed at for the stores illustrated in the text.

VAPOUR BARRIERS

There are several ways of measuring moisture vapour transmission rates but there are no clearly defined standards. For this reason, comparative figures for vapour diffusion are not included and manufacturers' advice should be sought in difficult cases. The materials listed below are those commonly used in the construction of cool stores.

Rigid barriers   Steel sheet, aluminium sheet and selected plastic sheets. When fixed securely and all joints and holes completely sealed, these provide a virtually impermeable barrier.

Flexible membranes   Aluminium foils are the most effective and can give almost complete resistance. Coated papers, plastic films such as polythene and PVC give adequate resistance for most purposes, but they must be lapped, sealed and unpunctured.

Applied coatings   Bitumen based proprietary formulations applied by brush or trowelling are effective. Resin based and other special applications may be suitable if continuous and applied according to the manufacturers' instructions.

Appendix 2c

INSULATION VALUE OF SOME COMMONLY USED MATERIALS*

Material	Density (kg/m³)	k Value (W/m°C)
Traditional building materials		
Bricks	1762	0·808
Concrete (ballast)	2400	1·440
Timber (deal)	609	0·125
Blocks (insulating)	800	0·200
Boards, sheets and loose-fill		
Asbestos insulating board	720	0·108
Cork board	110	0·038
Regranulated cork (loose-fill)	80	0·036
Expanded polystyrene	25	0·033
Expanded polystyrene granules (loose-fill)	16	0·035
Glass fibre	50	0·033
Glass fibre (loose-fill)	80	0·036
Mineral wool	50	0·036
Mineral wool (loose-fill)	128	0·036
Polyurethane board	30	0·023
Wood wool	400	0·084
Foamed 'in situ'		
Urea formaldehyde	8	0·036
Polyurethane	40	0·024
Sprayed-on		
Polyurethane	40	0·023

*The figures given are only approximate and reference should be made to manufacturers' specifications for greater accuracy.

Appendix 2d

$(U = W/m^2\ °C)$

U value not more than:

Store	Operating temperature	Ceilings and external walls	Roof	Floor	Insulation method illustrated
A	3·3°C October–March	0·43	—	0·40	2 layers of expanded polystyrene to walls and ceiling to total thickness 80 mm. 1 layer corkboard to floor 40 mm thick.
B	0°C throughout year	0·34	—	0·40	Expanded polystyrene 100 mm thick in wall and ceiling panels. 1 layer of high density expanded polystyrene to floor 40 mm thick.
C	2°C Intermittent short-term throughout year	0·43	—	0·70	2 layers of expanded polystyrene to walls and ceiling to total thickness 80 mm. Vertical strip of expanded plastic 50 mm thick, 450 mm deep to floor perimeter.
D	3°C October–March	0·46	0·50	—	Polyurethane sprayed-on insulation 50 mm thick to walls and roof.
E	0°C throughout year	0·34	—	0·40	Expanded polystyrene 100 mm thick in wall and ceiling panels. 1 layer corkboard to floor 40 mm thick.
F	2°C October–June	0·46	—	—	Polyurethane 50 mm thick in internal lining to walls and ceiling.

*Figures given as a guide only. Each store must be designed to suit the particular conditions and performance required.

Appendix 3a

RELATIVE HUMIDITY TABLE FOR USE WITH A SLING WET AND DRY BULB HYBROMETER

Dry bulb reading (°C)	Depression of wet bulb (°C)														
	0·5	1·0	1·5	2·0	2·5	3·0	3·5	4·0	4·5	5·0	5·5	6·0	6·5	7·0	7·5
35	96	93	90	87	84	80	77	74	71	68	66	63	61	58	55
34	96	92	89	86	83	80	77	74	70	68	65	62	60	57	54
33	96	92	88	86	83	80	76	73	70	67	64	62	59	56	54
32	96	92	88	86	83	80	76	73	70	67	64	61	58	56	52
31	96	92	88	85	82	79	76	72	69	66	64	60	58	55	52
30	96	92	88	85	82	79	76	72	69	66	63	60	57	54	51
29	96	92	88	85	82	78	75	71	68	65	62	59	56	54	51
28	96	92	88	84	81	78	74	71	68	65	62	58	56	53	50
27	96	92	88	84	81	77	74	70	66	64	61	58	55	52	49
26	96	92	88	84	80	77	74	70	66	64	60	57	54	51	48
25	96	92	88	84	80	76	73	70	66	63	59	56	53	50	47
24	96	92	88	84	80	76	72	69	65	62	58	55	52	49	46
23	96	91	87	84	80	76	72	68	64	61	58	54	51	48	44
22	96	91	87	83	79	75	71	67	64	60	56	53	50	46	43
21	96	91	87	83	79	75	70	67	63	59	55	52	48	45	42
20	95	91	87	83	78	74	70	66	62	58	54	51	47	44	40
19	95	91	86	82	78	73	69	65	61	57	53	50	46	42	38
18	95	91	86	81	77	72	68	64	60	56	52	48	44	40	37
17	95	90	85	81	76	72	67	63	59	55	50	46	43	39	35
16	95	90	85	81	76	71	66	62	58	53	49	45	41	37	33
15	95	90	85	80	75	70	65	61	56	52	48	43	39	35	31
14	95	90	85	79	74	69	64	60	55	50	46	42	37	33	29
13	95	90	84	78	73	68	63	58	54	49	44	39	35	31	26
12	94	89	84	78	72	67	62	57	52	47	42	38	33	28	24
11	94	89	83	77	71	66	61	55	50	45	40	35	30	26	21
10	94	89	82	76	70	65	59	54	48	43	38	33	28	23	18
9	94	89	81	75	69	64	58	52	47	41	36	30	25	20	15
8	93	88	81	74	68	62	56	50	45	39	33	28	23	17	—
7	93	88	80	73	67	61	54	48	42	36	31	25	19	—	—
6	93	86	79	72	66	59	53	46	40	34	28	—	—	—	—
5	92	85	78	71	64	58	51	44	38	—	—	—	—	—	—
4	92	84	77	70	63	56	48	—	—	—	—	—	—	—	—
3	92	84	76	68	63	—	—	—	—	—	—	—	—	—	—
2	92	84	76	—	—	—	—	—	—	—	—	—	—	—	—
1	92	83	—	—	—	—	—	—	—	—	—	—	—	—	—

To find the relative humidity, select the dry bulb temperature in the left hand column. Knowing the difference between the dry and wet bulb (depression of wet bulb), read down the appropriate column until this meets the dry bulb line across the page. The figure given is the per cent relative humidity.

Appendix 3b

DEWPOINT TABLE

Temperature of air °C

RH	0	2	4	6	8	10	12	14	16	18	20	22	24	26
90	−1·3	0·5	2·4	4·5	6·5	8·5	10·5	12·4	14·4	16·4	18·4	20·4	22·3	24·3
85	−1·9	−0·2	1·8	3·7	5·7	7·7	9·6	11·6	13·5	15·5	17·5	19·4	21·4	23·4
80	−2·7	−0·9	0·9	2·8	4·8	6·7	8·7	10·7	12·6	14·6	16·5	18·5	20·4	22·4
75	−3·4	−1·7	0·0	1·9	3·9	5·6	7·6	9·7	11·6	13·6	15·5	17·5	19·4	21·4
70	−4·2	−2·5	−0·8	1·0	2·9	4·8	6·8	8·7	10·6	12·5	14·5	16·4	18·3	20·3
65	−5·0	−3·4	−1·7	0·0	1·9	3·8	5·7	7·6	9·5	11·5	13·4	15·3	17·2	19·1
60	−6·0	−4·3	−2·7	−1·0	0·8	2·7	4·6	6·5	8·4	10·3	12·1	14·0	15·9	17·9
50	−8·1	−6·5	−4·8	−3·2	−1·5	0·1	2·0	3·9	5·7	7·6	9·4	11·3	13·2	15·0

The temperatures in the table are those at which water vapour will start to condense. For example, with air at 20°C and 70 per cent RH, water will condense on surfaces at or below 14·5°C.

K

Appendix 4a

STORAGE DESIGN: RECOMMENDED CONDITIONS

	Temperature (°C)	Relative Humidity (per cent)
Onions—Bulb	minus 1–0	70–80
Asparagus, Broad beans, Cabbage, Carrots, Cauliflower, Celery, Leeks, Lettuce, Onions —salad, Parsnips, Peas, Rhubarb, Spinach, Sprouts, Sweet Corn, Water Cress, Soft Fruit (strawberries, raspberries, etc.)	0–2	over 95
Red Beet	3	over 95
Beans—runner and dwarf	4–7	over 95
Potatoes (canning)		
Potatoes (main crop)	7 (sprout suppressant necessary)	90–95
Tomatoes	8	85
Cucumber, melons, peppers, courgettes, marrows	7–10	over 95
Potatoes (for crisping)	10 (sprout suppressant necessary)	over 95

Readers are also referred to the notes under the crop headings in the chapter on storage of vegetables.

Storage requirements of *apples* and *pears* are more precise. Details are given in Tables 1 to 3 on pages 3–6.

Appendix 4b

HEAT PRODUCTION DUE TO RESPIRATION IN AIR (MAXIMUM VALUES)

Temperature (°C)	Heat production (W/tonne)					Specific heat (kJ/kg °C)	Ref.
	0	5	10	15	20		
Fruit							
Apples	10	19	30	39	46	3·64	2
Blackberries	63	94	177	214	444	3·68	1
Black currants	45	77	111	257	372	3·68	1
Cherries, sweet	16	47	91	130	160	3·52	3
Grapefruit	—	19	30	48	78	3·81	5
Gooseberries	26	49	75	92	100	3·81	2
Oranges	14	23	38	63	103	3·77	5
Pears	11	39	73	110	156	3·60	2
Plums	18	36	63	105	165	3·73	2
Raspberries	69	158	92	389	576	3·56	1
Rhubarb, forced	40	60	100	126	155	4·02	1
Strawberries	43	80	147	245	374	3·85	1
Strawberry plants	34	—	—	—	—	—	10
Tomatoes	17	26	43	66	86	3·98	1
Vegetables							
Artichokes	100	140	212	330	533	3·64	4
Asparagus	80	126	180	300	363	3·94	1
Beans, Broad	104	155	259	357	432	—	1
Beans, Runner	63	83	107	161	269	3·81	1
Brussels Sprouts	51	89	149	223	268	3·68	1
Cabbage, Primo	33	77	89	110	119	3·94	1
Cabbage, January King	18	39	48	98	170	3·94	1
Cabbage, Winter White	9	21	24	39	60	3·94	1
Calabrese	125	173	313	596	715	—	1
Carrots (without tops)	39	51	57	71	98	3·81	1
Carrots (immature with tops)	104	152	220	316	360	3·81	1
Cauliflower	60	101	134	199	375	3·89	1
Celery	21	27	36	44	98	3·98	1
Cucumber	17	22	37	40	42	4·06	1
Green Peppers	23	31	57	63	100	3·94	1
Leeks	60	83	149	223	328	3·68	1
Lettuce, cabbage	48	71	92	149	238	4·02	1
Mushrooms	130	210	348	570	930	3·89	6
Onion, dry bulb	9	15	21	21	24	3·77	1
Onion, salad	54	105	176	276	410	3·81	2
Parsnip	21	33	77	98	146	3·48	1
Peas (in pod)	140	164	357	506	744	3·30	1
Peas (shelled)	217	290	460	1070	1600	3·31	7
Potato, main crop	18	9	12	15	18	3·43	1
Potato, early	30	45	60	89	119	3·56	1
Radish (with tops)	49	62	107	207	405	4·02	8
Radish (without tops)	27	40	65	109	183	4·02	8
Red Beet (without tops)	12	21	33	51	57	3·77	1
Red Beet (with tops)	33	42	66	74	119	3·77	1
Spinach	149	208	238	357	447	3·94	1
Sprouting broccoli	229	357	506	819	1265	—	1
Sweet corn	89	158	259	409	605	3·31	1
Turnip (with leaves)	45	51	89	128	155	3·89	1
Turnip (without leaves)	19	40	57	65	69	3·89	2
Watercress	52	104	230	392	597	3·98	1
Flowers							
Carnation	28	45	86	190	690	—	2
Narcissus	74	140	246	410	620	—	11
Rose bushes	43	—	—	—	—	—	9

Appendix 4c

CARBON DIOXIDE PRODUCTION (g/TONNE h) IN AIR AND CA CONDITIONS

| | Temperature, °C | | | | | | |
	0	3·5	5	10	15	20	Ref.
Apples							
Air			11–10	16–15	27–22	45–32	12
2% O_2		6					13
5% CO_2 3% O_2		3					13
8% CO_2 13% O_2		3					13
5% CO_2 16% O_2		5					13
Pears							
Air			14	24	33	39	12
Strawberries							
Air	15			52		127	1
3% O_2	12			45		86	1
Raspberries							
Air	24			92		200	1
3% O_2	22			56		130	1
Cauliflower							
Air	20			45		126	1
3% O_2	14			45		60	1
Storage cabbage							
Air	3			8		20	1
3% O_2	2			6		12	1

Appendix 4d

REFERENCES TO APPENDICES IV (B AND C)

1. Robinson, J E, Browne, K M and Burton, W G. Storage characteristics of some vegetables and soft fruits. *Ann. appl. Biol.*, **81**, 399–408 (1973).

2. Smith, W H. Lfl. Fd. Invest. Bd., **15**, 7 pp. (1952).

3. English, H and Smith, E. Respiration, internal atmosphere and moisture studies of sweet cherries during storage. *Proc. Am. Soc. hort. Sci.*, **41**, 119–123 (1942).

4. Rapport, L and Watada, A E. Effect of temperature on artichoke quality. *Proc. Conf. Transportation Perishables*, Davis, Calif. 142–146 (Feb. 1958).

5. Harding, P L, Lutz, J M and Rose, D H. The respiration of some fruits in relation to temperature. *Proc. Soc. hort. Sci.*, **28**, 583–589 (1931).

6. Smith, W H. The storage of mushrooms. *Rep. Ditton and Covent Gdn. Lab. for 1963/64*, -- 18 (1964).

7. Tewfik, S and Scott, L E. Respiration of vegetables as affected by post harvest treatment. *J. agric. Fd. Chem.*, **2**, 415–417 (1954).

8. Platenius, H. Effect of temperature on the rate of deterioration of fresh vegetables. *J. agr c. Res.*, **59**, 41–58 (1939).

9. Nahlstede, J P. Shipping roses in polyethylene wraps. *Am. Nurserym.*, **101 (2) 7**, 92–98 (1955).

10. Lockhart, C L and Eaves, C A. The influence of controlled atmospheres on the storage of strawberry plants. *Can. J. Pl. Sci.*, **42**, 151–145 (1966).

11. Post, K and Fischer, C W. Commercial storage of cut flowers. *Ext. Bull. Cornell agric. Exp. Stn.*, **853**, 14 pp.

12. Wilkinson, B G. The effect of temperature on the production of ethylene by apples and pears. *Rep. E. Malling Res. Stn. for 1972*, 118–120 (1973).

13. Sharples, R O. E. Malling Res. Stn. *Private Communication* (1978).

Appendix 4e

PRODUCE DENSITY

Product	Product only (m³/tonne)	Typical stored volume* (m³/tonne)
Fruit		
Apples	1·90	3·70 (b)
		4·40 (bb)
Cherries, sweet	2·46	
Gooseberries	2·05	
Pears	1·59	3·10 (b)
Strawberries	2·18	8·81 (t)
Tomatoes	2·44	5·92 (c)
Vegetables		
Cabbage, Savoy	3·57–4·18	
Cabbage, Spring	6·3	
Cabbage, Winter White	2·55	4·07 (b)
Carrots	1·70–1·84	2·63 (b)
Cauliflower	2·25–4·50	
Lettuce	4·03–5·68	10·59 (c)
Onions	1·8–2·83	3·85 (b)
Peppers	4·88	11·84 (c)
Potatoes	1·42–1·59	2·26–2·54 (b)
Red Beet (baby)	1·46	
Red Beet (Mature)	1·70	2·69–3·00 (b)
Sprouts	2·27	

*Actual spacial requirements will vary with type of container and method of stacking. If the product density is known, store volume may be estimated by multiplying by 1·5 if bulk bins are used.

  (b)=bulk bins
  (c)=cartons
  (t)=trays
bb)=bushel boxes

Appendix 4f

SUGGESTED DESIGN COOLING TIMES

Strawberries and Raspberries Sweet Corn Asparagus Calabrese Spinach	3 hours
Runner Beans Lettuce Spring Onion Spring Cabbage Cauliflower Sprouts	8 to 12 hours
Carrots (Immature) Leeks Tomatoes	12 to 18 hours
Carrots (Mature) Apples Pears	4 to 7 days
Red Beet Winter White Cabbage	14 to 20 days
Potatoes (Mature) Onions	3 to 6 weeks

COOLING TIMES (h) OF SOFT FRUIT AT VARYING TEMPERATURES WITH DIFFERENT PLANT CAPACITIES

Fruit temperature °C	Refrigerating capacity (kW/tonne fruit)									
	1·0	3·0	5·0	7·0	9·0	11·0	13·0	15·0	17·0	19·0
27·0	0·0	0·0	0·0	0·0	0·0	0·0	0·0	0·0	0·0	0·0
26·0	1·1	0·4	0·2	0·2	0·1	0·1	0·1	0·1	0·1	0·1
25·0	2·2	0·7	0·4	0·3	0·2	0·2	0·2	0·1	0·1	0·1
24·0	3·2	1·1	0·6	0·5	0·4	0·3	0·2	0·2	0·2	0·2
23·0	4·3	1·4	0·9	0·6	0·5	0·4	0·3	0·3	0·3	0·2
22·0	5·4	1·8	1·1	0·8	0·6	0·5	0·4	0·4	0·3	0·3
21·0	6·5	2·2	1·3	0·9	0·7	0·6	0·5	0·4	0·4	0·3
20·0	7·6	2·5	1·5	1·1	0·8	0·7	0·6	0·5	0·4	0·4
19·0	8·7	2·9	1·7	1·2	1·0	0·8	0·7	0·6	0·5	0·5
18·0	9·7	3·2	1·9	1·4	1·1	0·9	0·7	0·6	0·6	0·5
17·0	10·8	3·6	2·2	1·5	1·2	1·0	0·8	0·7	0·6	0·6
16·0	11·9	4·0	2·4	1·7	1·3	1·1	0·9	0·8	0·7	0·6
15·0	13·0	4·3	2·6	1·9	1·4	1·2	1·0	0·9	0·8	0·7
14·0	14·1	4·7	2·8	2·0	1·6	1·3	1·1	0·9	0·8	0·8
13·0	15·2	5·1	3·0	2·2	1·7	1·4	1·2	1·0	0·9	0·9
12·0	16·2	5·4	3·2	2·3	1·8	1·5	1·3	1·1	1·0	0·9
11·0	17·3	5·8	3·5	2·5	1·9	1·6	1·4	1·2	1·1	1·0
10·0	18·4	6·1	3·7	2·6	2·0	1·7	1·5	1·3	1·2	1·2
9·0	19·5	6·5	3·9	2·8	2·2	1·8	1·6	1·4	1·3	1·3
8·0	20·6	6·9	4·1	2·9	2·3	2·0	1·7	1·6	1·5	1·4
7·0	21·7	7·2	4·3	3·1	2·5	2·1	1·9	1·8	1·7	1·6
6·0	22·7	7·6	4·6	3·3	2·7	2·3	2·1	2·0	1·9	1·8
5·0	23·8	7·9	4·8	3·6	3·0	2·6	2·4	2·2	2·1	2·1
4·0	24·9	8·3	5·2	4·0	3·3	3·0	2·7	2·6	2·5	2·4
3·0	26·0	9·0	5·8	4·6	4·0	3·6	3·4	3·2	3·1	3·1

Plant capacity limits cooling

Produce heat transfer limits cooling

Agricultural Chemicals Approval Scheme

In order to choose a proprietary product the reader should consult the list of *Approved Products for Farmers and Growers* issued under the Ministry's Agricultural Chemicals Approval Scheme and available in February of each year as a priced publication from booksellers or from HMSO, PO Box 569, London SE1 9NH. The label on the product gives detailed instructions on the dose to use, the dilution and the stage of growth of crop when spraying may be carried out. The container of an approved product bears the mark shown here. It is strongly recommended that approved products should be used.

Care with chemicals

Whenever chemicals are used, FOLLOW THE INSTRUCTIONS ON THE LABEL.

The effects of chemicals on different varieties, particularly of apples, may differ; the label will clarify the position.

Read:	— Some advice on the safe use of agricultural chemicals at the beginning of *Approved Products for Farmers and Growers*.
Consult:	Chapters in *Approved Products for Farmers and Growers* on
	— Health and Safety (Agriculture) (Poisonous Substances) Regulations.
	— Chemicals subject to the Poisons rules.
	— First Aid Measures.
	— Application of Agricultural Chemicals.

Protection of users   The Health and Safety at Work etc. Act 1974 imposes obligations on employers, the self-employed and employees who work with any chemicals.

Read: — *Take care when you spray.*
Paraquat is subject to the Poisons Act 1972. It is only available to *bona fide* farmers and growers who have to sign the poisons register on purchase. Organo-mercury compounds, Mercuric Oxide and 2-phenyl-phenol are also subject to this Poisons Act.

Protection of consumers   Allow at least the minimum intervals, as stated on the label, between the last application of chemicals and harvesting.

Dichlofluanid	2 weeks (strawberries and raspberries).
	3 weeks (blackberries, currants, gooseberries, loganberries).
Thiram	2 weeks (lettuce).
	1 week (other edible crops).

Protection of the environment   To safeguard bees, do not spray crops or weeds in flower.

Read: —*Code of practice for the disposal of unwanted pesticides and containers on farms and holdings.*
Dispose of empty containers safely.
Do not contaminate ponds, ditches or waterways.
Store new and part used containers in a secure place.

All the leaflets referred to in this book are available from MAFF Regional and Divisional Offices or from the address on page 67.

Index

Thiabendazole 35, 37
Thiophanate-methyl 35, 36, 37, 40–1, 42
Thiram 15, 40, 144
Tip burn 62
Tomatoes 46, 53, 136, 137, 140
Transport 20, 23, 47
Trichothecium sp. 38
Turnips 137
Turnip mosaic virus 62

U values 77, 130, 132

Vacuum cooling 66, 87
Vapour barrier 78, 131
Vehicles—*see* transport
Venturia sp. 38
Vermin proofing 82
Virus, effect of 62

Wall structure 74
Water core 7, 33
Watercress 136, 137
Weather 2, 7, 28, 45
Wet bulb, depression of 134

Printed in Scotland by Her Majesty's Stationery Office at HMSO Press, Edinburgh
Dd 587014 K23 12/79 (16227)